Houseplants

Houseplants

Essential know-how and expert advice for gardening success

CONTENTS

PLANNING YOUR PLANTING

Before heading out to buy houseplants, it's important to give some thought to the different ways they can be positioned and displayed so that they'll not only look great but will also thrive. This section is packed with inspiration and advice on which rooms to put your plants in, how best to place and group them, and how to use their texture, colour, shape, and size to create a specific mood or style. Whatever your vision, with some good planning, plants can transform your home and enrich your life.

WHY GROW HOUSEPLANTS?

Contact with nature has been proven to relieve stress and improve wellbeing, and has positive and measurable effects on physical health too. Houseplants, with their diverse forms and colours, deliver these benefits directly into your home, whether you live in a mansion or an urban flat. Looking after plants not only enlivens your house and gives expression to your creativity but is hugely rewarding, satisying a deep-seated need to nurture and an instinct to make life better.

MAKE FRIENDS

The simplest reason to grow plants in your home is that they are upliftingly beautiful. Seeing greenery around you can boost your mood, make you feel more relaxed and calm, and provide a number of health benefits (see pp.10–11). You don't need a lot of space to fit in some stunning specimens, and as you read this book, you'll soon see that there's a houseplant for every room and every occasion. Each houseplant has its own character and, given time, will become your friend. Research suggests that talking to plants really does make them grow faster, too.

Trailing houseplants are cleverly arranged here to accentuate the height and lightness of an urban interior.

The false shamrock has distinctive purple leaves that close up at night.

CREATE AN ATMOSPHERE

Houseplants can transform your living space immeasurably. Whether your tastes lean towards the minimal or the grand and baroque, there's sure to be a species – or even a creative combination of plants – with the aesthetic attributes to create the style and atmosphere you're aiming for. The bold, architectural forms of cacti, the stunning flowers of amaryllis and hibiscus, and the lush tropical foliage of plants such as palms, bromeliads, and philodendrons – all create their distinctive moods and can give one space very different looks.

Houseplants can be used to unite or divide interior spaces, to create patterns, to disguise features you'd rather lose, and to create interest in dead spaces. Their colours can complement or contrast with interior finishes and decorations to give you a new look in just a few minutes, without lifting a paint brush or investing in costly new furniture. Choosing and positioning houseplants, along with the containers you grow them in, presents you with countless options and ways to express your personality in your home; and what's more, if you ever move house, your garden can move with you.

DEVELOP AN INTEREST

An interest in houseplants often has small beginnings, perhaps triggered by the receipt of an unusual plant as a gift or by experiencing stunning foliage, flowers, and scents on a holiday to warmer climes. However your interest arises, it can soon take you in many directions. You may, for example, become captivated by one type of plant – orchids or carnivorous plants, perhaps – and be driven to collect and cultivate a range of such species.

You might also be inspired to grow plants from one particular country, continent, or ecosystem – cacti from American deserts or Brazilian bromeliads, for example. There's no need to have a theme in mind when keeping houseplants, but doing so turns plant collecting into a compelling pastime, through which you can build your appreciation of the living world.

Carnivorous plants provide a talking point in any household.

Inspiration for groupings may come from seeing plants in their natural habitats.

LEARN TO NURTURE

One of the most satisfying aspects of keeping houseplants is watching them mature into healthy specimens, and perhaps even flower and produce offspring. Many people who grow them are inspired to try outdoor gardening when they see what is possible with limited resources indoors.

When you begin keeping houseplants, select species that are relatively easy to grow and maintain, and that are well matched to the growing conditions in your home. Success will encourage you to move on to species that require more specialized care. Indoor gardening is an addictive hobby and before you know it, you'll have a plant in every room of the house, and a network of fellow enthusiasts with whom you can share knowledge and cuttings.

Propagating and repotting houseplants to share with friends and family is a terrific social activity.

PLANTS FOR A HEALTHY HOME

On average, we spend more than 85 per cent of our time indoors, often surrounded by technology and cut off from nature. What's more, our homes – once sanctuaries – are increasingly becoming places of work. Houseplants offset some of the stresses of modern working and living by providing closer contact with nature – a factor that has positive effects on our psychological and physical health.

As well as bringing joy and lifting our spirits, studies show that plants may also reduce stress and purify the air.

BOOSTING WELLBEING

Houseplants do much more than enhance the look of your home. Studies have shown that they deliver a wide range of health benefits too. Their presence helps to elevate mood, reduce stress and fatigue, and even lower blood pressure and heart rate. Patients in hospital are known to recover faster, and report lower levels of pain, when surrounded by plants. The mechanisms underlying these responses are still being studied, but are in part linked to the associations humans make with different colours. Green is linked closely with feelings of refreshment, positivity, and peace – it's no accident, for example, that guests on TV shows relax in "the green room" and that doctors' waiting rooms are often decorated in the colour.

Handling and looking after plants promotes a positive outlook on life, and these activities are fundamental to programmes of "horticultural therapy" that have been developed by psychologists to help people overcome feelings of loneliness, anxiety, and depression.

Creating a green space by filling a room with plants can help you establish a sanctuary for meditation or yoga.

AIR CONDITIONING

In the late 1980s, NASA commissioned research into the use of plants as a means of purifying the air in space stations. They revealed that some plants were effective in removing volatile organic pollutants such as benzene, formaldehyde, and tricholoethane – substances present in, or used to make, a number of household products.

How well this laboratory research applies to the home environment is a matter for continuing investigation. However, the presence of a wide range of plants in your living space is certainly effective in increasing humidity by up to 20 per cent, which can be of benefit to respiratory and skin health, particularly in the winter months.

Plants credited with air-purifying qualities include aloe vera (see p.53), parlour palm (see p.64), spider plant (see p.65), and dragon plant (see p.80).

Plants help create healthier homes by regulating temperature and humidity.

GROW PRODUCTIVE

A study in 2019 carried out by Dutch scientists found a reduction of 20 per cent in sick days taken by people who work in plant-filled offices. The same employees also reported an uplift in their mood and greater satisfaction in their work performance.

Environmental psychologists also suggest that the presence of plants in an interior space makes people more attentive, improves their reaction times, and can nourish creativity. Theorists speculate that proximity to plants satisfies an instinctive desire to be close to food sources, which in turn reduces anxiety and allows us to devote our brain power to more creative tasks.

Biophilic design – creating workplaces that reinforce the connection between people and nature – is increasingly becoming big business.

TOP TIP SURROUND YOURSELF WITH A WIDE RANGE OF HOUSEPLANTS WHEN YOU WORK AND, OVER TIME, YOU'LL BEGIN TO FEEL MORE COMFORTABLE AND CALM, AND DEVELOP GREATER FOCUS AND CREATIVITY.

A GATEWAY TO GARDENING

Growing houseplants is fun, creative, and enriching. Many accomplished outside gardeners credit the start of their lifelong gardening passion to a houseplant. It's common for a pocket-money plant such as a Venus flytrap to get young growers hooked. Others often pick up the hobby when they move house and realize how just a few plants can completely transform the style and mood of their environment.

Succulents are a terrific choice for beginner gardeners as they are easy to care for and rarely need watering.

Herbs thrive on a sunny windowsill – the secret is to pick them regularly and grow the ones you use most often in cooking.

SURE STARTERS

Buying easy-to-grow plants is a great way to encourage new gardeners to enjoy success. Here are some starter plants that are virtually bullet-proof and almost impossible to kill:

TOP STARTER PLANTS Aloe vera (*Aloe vera*, see p.53) • Living stone (*Lithops*, see p.106) • Money plant (*Crassula ovata*, see p.71) • Parlour palm (*Chamaedorea elegans*, see p.64) • Spider plant (*Chlorophytum comosum*, see p.65) • Sweetheart plant (*Philodendron scandens*, see p.120) • Umbrella tree (*Schefflera arboricola*, see p.130)

The spider plant has been popular for decades as it's so easy to grow.

CONFIDENCE BUILDERS

New gardeners are often given their first houseplant as a gift and this sparks the beginning of what can become a lifelong interest. Among the many attractions of indoor gardening is that there's no digging or weeding involved: houseplants are easy to care for, display, and admire, so those who are new to the hobby will soon experience the immense pleasure that nurturing a houseplant inevitably brings.

Success with houseplants leads to confidence in growing them and a desire to tend plants that live both indoors and out. The next step for most people is to grow edible herbs on the windowsill and, if they have the luxury of a garden, seasonal container displays often follow.

GETTING CLOSER TO NATURE

It's very easy to get attached to a houseplant. Living with them every day, watching them grow and flower, gives indoor gardeners direct and immediate appreciation of nature. Gardeners quickly begin to realize how magical it is to nurture a plant and how easy it is to get results.

Those who haven't grown up with a garden will enjoy getting their hands dirty when potting on or taking cuttings. This complete involvement in a plant's life-cycle encourages a deeper interest in, and connection with, the natural world (see *pp.10–11*). Once the shelves and windowsills in the home are filled with an ever-growing collection of houseplants, new gardeners will soon venture outside to widen their plant portfolio. Success with houseplants can often inspire those without gardens to sign up for an allotment.

Many beginner gardeners enjoy getting their hands dirty.

TOP TIP IF YOU HAVE A BALCONY OR PATIO, TRY MOVING YOUR FOLIAGE HOUSEPLANTS OUTSIDE DURING THE SUMMER. LIKE MANY NEW GARDENERS, YOU'LL SOON REALIZE THE IMPACT PLANTS CAN HAVE ON YOUR OUTSIDE SPACE. THE NEXT STEP IS TO INVEST IN HARDY PLANTS THAT OFFER YEAR-ROUND OUTSIDE INTEREST.

African violet cuttings are easy to take and soon root in water (see *pp.40–41*).

Encouraging hyacinths to bloom in time for Christmas is fun for everyone.

ENCOURAGING NEW GARDENERS

Houseplants can be used to generate conversation, fun, and entertainment in the home. For example, encourage household members to get involved in potting up plants or taking leaf cuttings (see *pp.38–41*). You can also interest them in other activities, such as propagating and growing plants in water so they can see just how rapid and effective the results can be (see *p.42*). Make a note of when cuttings, bulbs, or plants were placed in water and keep the plants in a room everyone uses so that they can track progress. Giving friends and family plants you've propagated yourself is another great way to spark their interest.

CHOOSING THE RIGHT LIGHT

For plants to be able to successfully perform photosynthesis – the process by which they produce fuel for growth – their leaves must absorb energy from sunlight. The amount of light they require for good health depends on where they originate – a tropical rainforest or a desert, for example. The north-, south-, east-, and west-facing rooms in a house all have different levels of light and offer different growing opportunities.

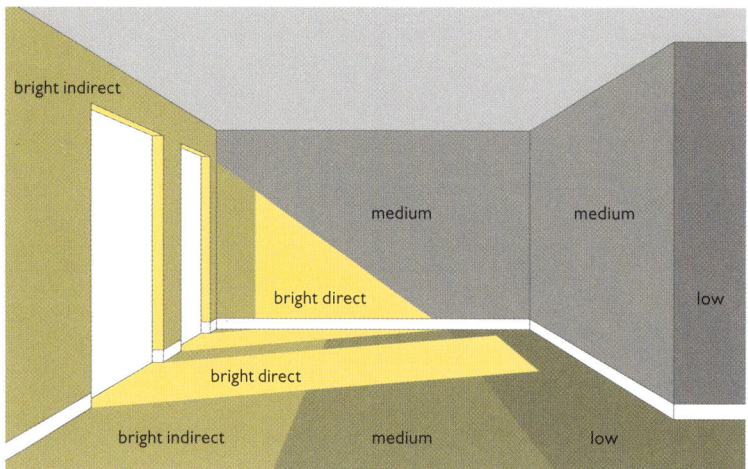

bright indirect

medium

medium

bright direct

low

bright direct

bright indirect

medium

low

ASSESSING LIGHT LEVELS

Observe your room throughout the day to work out how the light changes as the sun moves – you can use light meters to take readings. Assess which way the room faces as different aspects will receive different amounts of light. A room that's sunny in the morning can be shady by noon, and vice versa. The shape of a room also affects the amount of light reaching certain points.

Different growing zones and light levels exist in any one room. You can therefore often grow both sun- and shade-loving plants in the same space.

SOUTH-FACING ROOMS

A south-facing room with a large window is the dream for many indoor gardeners. These rooms offer plants the highest temperatures and light levels. With luck, sun will stream in during the day – an ideal setting for plants that come from arid countries.

Cacti and succulents – echeverias, for example (see p.83) – thrive in a south-facing room and its heat and light levels encourage flowering. However, plants that prefer shade and high humidity will suffer – their foliage is at risk from being scorched by the sun.

Plants can easily dry out in a south-facing room: be vigilant about watering and move them to a shadier spot if you're away for any length of time.

Succulents enjoy the heat and direct sunlight offered by a south-facing room.

WEST-FACING ROOMS

A west-facing aspect is a happy medium for most plants and the range you can grow here will be wide. The room will be filled with afternoon and evening sunlight, which has far more warmth than the morning sun offered by an east-facing room. West-facing rooms offer a great growing environment – it's almost worth giving any plant a try in this spot, and it's always a safe bet if you're unsure of a plant's needs.

This location is ideal for a collection of African violets (see p.126) – they need warmth and brightness to flower but their foliage would suffer from scorch if placed in a south-facing aspect.

African violets will flower profusely if displayed on a west-facing windowsill.

EAST-FACING ROOMS

This is the perfect place for plants that prefer filtered sunlight rather than direct sun for most of the day. An east-facing room will receive morning sun and then filtered sunlight in the afternoon. Plants such as the scarlet star (see p.94), which would naturally grow under the canopy of trees and plants in a tropical forest, will flourish in this location.

The variegated or colourful foliage of plants that enjoy filtered light can fade over time in this room due to lack of light in winter. Remedy this by moving them to a south- or west-facing aspect.

Plants can be misted in a north-facing room without fear of the sunlight scorching their foliage.

NORTH-FACING ROOMS

Although these are the darkest rooms in the house, many plants love the shady conditions. In summer, a north-facing room will receive some evening sun, but in winter it will be far darker. Move plants that are reaching for the light to a west-facing aspect in the colder, darker months of the year and return them to the north-facing aspect in summer.

The plants that flourish in this shady spot tend to be those with large, dark green foliage. The Chinese evergreen (see p.52) will grow happily here, as will the rubber plant (see p.88).

Be very careful not to overwater plants growing in a shady room as compost will take longer to dry out.

East-facing rooms provide indirect light in the afternoon for shade- and moisture-loving plants.

CHOOSING THE RIGHT ROOM

Choosing the right location for your houseplants is the surest way to guarantee success. Every home is configured differently, but there are some common factors that can help you to determine which room a plant is most suitable for. Humidity levels, draughts, temperature, and light (see pp.14–15), are all key factors in deciding the best place to display your plants.

To really thrive, houseplants need to feel as comfortable as you do.

A cool hallway is the perfect place for the highly scented paper white narcissus.

HALLWAYS

A display of houseplants in a hallway is a great welcome, but it can be a challenge to find the right plant for such a tricky area. Hallways are often short of natural light and, with doors constantly opening and closing, are subject to cold draughts and sudden changes of temperature. A draught excluder at the foot of the door can help prevent draughts at night.

Choose plants for this area that are robust and able to cope with these fluctuating conditions – the cast iron plant (see p.56) is a good contender. If the temperature in a hallway is low, this could be the perfect place to display seasonal plants such as the paper white narcissus (see p.110) as its white, scented flowers will last longer.

SITTING ROOMS AND BEDROOMS

Sitting rooms and bedrooms offer similar opportunities and growing environments for plants. The humidity levels tend to be low to moderate, so choose plants that will thrive in these conditions, such as the kentia palm (see p.100). If the rooms are small or poorly ventilated, avoid any with overpowering scents, such as hyacinths. Also avoid placing toxic plants or those with spines or sharp edges in children's bedrooms.

These are rooms in which people tend to be sedentary, so they're often well heated and the temperature can vary greatly during the day and night. The vast changes in temperature that plants undergo if they're near a heat source can cause sudden leaf drop, so avoid placing them close to radiators in winter. When outside temperatures drop and the heating goes on, it's time to consider moving the plants to a cooler part of the room. Draughts are rarely an issue in these rooms, and the light levels will depend on the size of windows and the aspect of the room (see pp.14–15).

When winter arrives, move your plants away from radiators.

BATHROOMS

Bathrooms generally have high levels of humidity, thanks to the steam from showers and baths. This environment replicates a hot and wet tropical forest and makes a perfect home for many jungle plants such as the bird's nest fern (see p.57), whose natural habitat is a tropical woodland. Bathrooms are often quite small rooms and therefore a popular choice for hanging-basket plants.

Many bathrooms have frosted glass windows, which provide plants with filtered rather than direct sunlight. Blinds are also often used for privacy in bathrooms, but it's important to lift these when the room isn't in use so that plants can receive natural light.

Bathrooms aren't walkthrough spaces and generally don't have a door leading to the outside, so it's unlikely plants will suffer any of the problems associated with draughts or sudden changes in temperature (see p.44–45).

Plants that flourish in a high-humidity environment are perfect for a steamy bathroom.

KITCHENS

Keeping a collection of houseplants on the kitchen windowsill makes perfect sense as they're positioned near the sink for watering and you can study them daily as they start to bloom. Humidity will be high to moderate in a kitchen due to the constant hot water from washing up and the steam from cooking. Cymbidiums (see p.75) are popular kitchen plants as they enjoy moderate humidity and have captivating flowers.

Kitchen surfaces can be easily swept clean and washed, so this tends to be the room where plants are potted up, cuttings taken, and offsets removed.

Unless there's a door to the outside, draughts are seldom a problem in this setting. However, care should be taken when growing plants on a windowsill as draughts and temperature extremes can be an issue here, particularly if the windows aren't double glazed. Also avoid placing houseplants close to a cooker as this too will subject them to dramatic changes of temperature.

As with all locations, light levels will be dependent on the style and aspect of the room (see pp.14–15).

NEED TO KNOW

- Protect furniture and carpets by placing a saucer under plant pots.
- Avoid putting plants close to electric points as this will make watering and misting hazardous.
- Pot plants can be heavy after watering, so ensure that shelving can take the weight.
- Place plants where they won't be knocked by passers-by.
- Never put plants next to open fires or on mantlepieces above them as their leaves will drop.

Keeping houseplants in the kitchen makes caring for them easier.

CREATIVE DISPLAYS

There are tremendous possibilities when it comes to displaying houseplants, however large or small. Plant them solo in a decorative container, or group them with other plants in bottle gardens, terrariums, or planters. Displaying plants in unusual and imaginative ways will give expression to your creativity, help define the character of your home, and bring out the very best in your plants.

Use imagination and creativity in your planting for maximum impact.

BOTTLE GARDENS

A bottle garden is an open-topped glass container that provides a microclimate for plants to grow in. A selection of small plants can be carefully placed in this container, where – if properly cared for – they'll thrive in the humid atmosphere. Bottle gardens are an ideal way of growing and displaying plants for anyone who is short of space or wants to create an impressive centrepiece.

A bottle or jar with a generous opening at the top is the easiest to plant up. To do this, place a layer of fine-grade gravel in the bottom of the jar for drainage. Top up with multipurpose compost mixed with a bit of fine charcoal to stop the compost from smelling. The compost should be deep enough to cover the plants' rootballs and allow for growth. Plant a selection of small, humidity-loving plants such as the polka dot plant (see p.102) and the radiator plant (see p.117). Choose plants

with contrasting colours, shapes, and textures to create a display that's rich in visual interest. If the opening of the jar is too small to reach inside, you may need to use chopsticks or specialist bottle-garden tools (see p.33).

Position out of direct sunlight to prevent the leaves from being scorched. Water with caution as the bottle will trap moisture and overwatering can be fatal. If one plant overshadows the others, prune it back to size.

Rub any lumps out of the compost before adding it to the jar in preparation for your miniature garden.

Remove the plants from their pots, then carefully tease out the roots before gently placing them in the bottle.

Choose plants that have interesting as well as contrasting textures and colours to achieve the most striking results.

TERRARIUMS

A terrarium is based on much the same principle as a bottle garden and allows small plants to be grouped together for decorative effect. However, terrariums are more elaborate in design and have a generous opening at the front. This means they don't become as humid inside as bottle gardens and therefore can be used for succulents (as well as for the plants used in bottle gardens).

To display succulents in this way, place a layer of gravel in the bottom of the terrarium to assist with drainage.

Add a layer of cactus compost with a little fine charcoal mixed in to keep it from smelling. Plant a selection of small succulents with contrasting leaf shapes, textures, and colours; position with the tallest at the back of the terrarium.

Water cautiously, allowing the compost to dry out between watering. Place so that the terrarium receives filtered sun. Most succulents enjoy bright sunlight, but the glass of the terrarium can cause the plants to overheat.

Some terrariums are designed both to hang and to sit on a tabletop.

Air plants don't need compost to grow and will thrive on just a piece of wood.

MOUNTING ONTO WOOD

Some epiphytic plants (those that grow on other plants) are well-suited to mounting onto pieces of decorative wood. This form of presentation mimics the plants' natural habitat and creates an impressive and unusual display.

If creating a display for a table, choose an attractive log or small tree stump; if mounting it on a wall, select a flatter piece of wood. Staghorn ferns (see p.123) are excellent candidates for this type of display. Pack around the edges of the plant's rootball with damp moss and then secure the moss and the rootball to the wood with a piece of clear fishing wire.

KOKEDAMA

There's no more dramatic way to display a houseplant than without a pot. The kokedama technique – whereby the roots of a plant are coated in moss and suspended in the air – is a great way to totally transform the look of small houseplants, such as maidenhair ferns (see p.50), that prefer light shade.

To make your own kokedama, simply mix equal parts of bonsai compost and multipurpose compost in a bucket and moisten. Remove your chosen plant from its container and take off any loose compost.

Pack the moist compost-mix firmly around the plant roots to form a ball and then cover this with a sheet of carpet moss. Wrap string around the neck of the plant to hold the moss in place and trim off any excess moss. Finally, hang your moss ball on a piece of string to display. Water the plant by submerging the rootball in water, then squeeze or drain off any excess water.

Display kokedama plants in a north-facing room to protect them from direct sun and stop them drying out.

TOP TIP THERE ARE NO RULES WHEN IT COMES TO CREATING HOUSEPLANT DISPLAYS. AS LONG AS THE PLANT IS FED, WATERED, AND CAN RECEIVE ENOUGH LIGHT, ALMOST ANY DISPLAY IDEA WILL WORK – EVEN HANGING PLANTS UPSIDE DOWN ISN'T OUT OF THE QUESTION.

DISPLAYING TRAILING AND CLIMBING PLANTS

Plants that clamber up supports or cascade magnificently from balconies or baskets add a huge amount of drama and spectacle to a room. This makes them great value plants and ideal for those who are short of floor or surface space. Decide how your plants are to be supported and trained and they'll be easy to control and sculpt to suit the exact requirements of the space.

Trailing plants allow you to fill a room with floor-to-ceiling greenery – a great way to inject vertical interest and colour.

A hanging display will weigh more when watered, so always make sure its support is adequate.

The style of your hanging basket can often add as much interest as the plant itself, so choose wisely.

HANGING BASKETS

Trailing plants introduce elegance and a different dimension to displays. There's a wide range to choose from, including the ever-popular spider plant (see p.65) and the silver inch plant (see p.139).

For those who are short of space, displaying in hanging baskets means you're able to fill an entire room with plants. There's a wealth of hanging containers available, from ceramic, wicker, and plastic to macramé and metal. Most are designed for a plant pot to be dropped into them, but if you're planning to plant directly into a hanging basket, make sure it's watertight. Choose a style that will suit your interior and be large enough for your chosen plant.

Large, heavy baskets must be hung with care. Seek the advice of a DIY expert before drilling holes in ceilings as some larger baskets will need to be hung from a joist. If you're not permitted to drill into ceilings, you can hang lighter baskets from clothes racks, banisters, hat stands, shelving, coat hooks or even curtain poles.

Watering baskets can be a challenge, so invest in a stepladder and a watering can with a long spout (see p.33).

> **TOP TIP** WHEN CREATING A MIXED PLANTER, PLACE YOUR TRAILING PLANTS AROUND THE INSIDE EDGE OF THE POT TO ENCOURAGE THEM TO TUMBLE OVER THE SIDE. THIS WILL NOT ONLY SOFTEN THE DISPLAY, BUT MAY ALSO HELP TO DISGUISE AN UNATTRACTIVE CONTAINER.

MOSS POLES

In their native environment, large plants with aerial roots, such as the Swiss cheese plant (see p.109), naturally use other plants as a frame to grow against. They don't have suckers to pull themselves up a support, but their aerial roots will hook onto a moist support and hold the plant upright. In a room setting, without help of this kind, plants become too sprawling for the available space and branches may be damaged under the weight of their own foliage.

To resolve this issue, some climbers and plants with aerial roots are trained up moss poles, which mimic the moist trunk of a jungle tree. These poles are widely available and can be bought ready-made in different heights. Simply push the pole into the plant pot and then use wire or string to attach the plant to the support. Keep the moss pole moist with regular misting. Over time, the aerial roots will gradually hook themselves onto it. You can easily prune out excessive growth to keep the plant the same height as the chosen moss pole.

Without a support, a Swiss cheese plant will outgrow its space.

LIVING SCREENS AND WALL COVERINGS

Climbers are a great way of covering a wall or creating a living screen in a room. Galvanized wires attached to a wall are the most inconspicuous supports available. Plants that have suckers, such as ivy (see p.97), will mark some walls, but you can avoid this by training them up free-standing supports. Trellis panels set away from a wall are a popular

choice. When securely attached to the back of a long trough, they make the perfect room divide.

Tie in climbers to a support to get them started, but as they mature, most will make their own way up it.

The biggest challenge comes in repotting. A plant that grows up a wall or trellis will be hard to remove from a pot after a few years. Instead of repotting, it's best to grow it in a large pot and then, each year, remove loose compost and replace it with a fresh layer.

The paper flower will remain a compact plant if trained appropriately.

TRAINING OVER HOOPS

Not everyone has the space for climbers to leap up walls and trellises. You can keep climbing and trailing plants confined to one easily movable pot by simply winding their stems around a hoop or tripod. Many plants will already be trained over a hoop when you buy them.

Lightly prune the plant throughout the year to keep it to the shape of the hoop. Be careful not to remove all the growth that will hold flowers.

Plants that respond well to this very straightforward form of training include the paper flower (see p.61) and stephanotis (see p.134).

Place climbing plants in large pots if you're growing them up walls. You can fix inconspicuous wires to the wall to help them clamber up.

INTERIOR DESIGN WITH PLANTS

The look of a room can be completely transformed by houseplants. They can be used to set a mood and help to create anything from a festive vibe to a relaxing retreat. When selecting, consider how their colour, scale, shape, and texture will interact with existing furnishings and colour schemes. There are no set formulas, but a few simple tips can make placing your plants less of a hit-and-miss affair.

One large statement plant can help to make a room feel bigger and create great impact.

COLOUR

Colour is a powerful element in any interior and can be used to evoke a mood, create coolness or warmth, and emphasize simplicity or opulence.

For a calming effect, choose silver and green foliage coupled with pale purple and white flowering plants. These cool colours will also help to make a room feel larger and brighter. For a more intimate or dramatic feel, opt for strongly variegated plants such as the croton (see p.70) and the blushing bromeliad (see p.111). Grouping plants with bright, clashing colours will create a vibrant high-energy display, whereas combining plants from a similar colour range or palette can be used to evoke a sense of order and control.

The containers you display plants in will greatly affect the look of the plants themselves and the impact they have on a room, so choose with care (see p.28).

Ferns are versatile plants that offset both modern and traditional interiors.

HOUSEPLANT TRENDS

Plants go in and out of fashion all the time, but don't let current trends restrict your creativity. You can alter the look of a plant by displaying it in an unexpected way. Ferns, for example, are often grown in hanging baskets, but if you grow three of them in identical pots and place them on a shelf or table, this retro plant can suddenly create a more sleek, refined look (see above).

Add a warm and intimate feel to a room with brightly coloured houseplants.

Identical plants in a neutral colour can create a modern, formal display.

SCALE AND SHAPE

With a wealth of houseplants available, you can have great fun experimenting with different shapes and sizes of plants when planning your displays.

The size of a room will to some extent restrict the size of plant you choose, but bear in mind that a small room will appear larger if coupled with one large, dramatic houseplant rather than lots of little plants.

If you're keen to evoke an intimate, cosy atmosphere in an interior, choose a contrasting combination of mounding, rosette-forming, upright, and trailing plants. If you'd prefer a more sleek, symmetrical, or formal look, then grouping plants of identical height and shape will achieve the desired effect.

Climbing plants trained up a trellis or large planters can be used to divide rooms and make more intimate spaces and areas. They can also be used very effectively to break up large expanses of bare wall. Screens of this kind are often used in open-plan offices.

Create a bold, symmetrical display by grouping plants with similar shapes.

TEXTURE

Plant textures have both visual and tactile appeal. Foliage can be shiny, velvety, prickly, or rough, and these different textures interact differently with light. Silky smooth and shiny leaves, for example, work well with room lighting in the evening.

Have fun matching or contrasting plant textures with the features in an interior. Velvety or feathery foliage will introduce a softer element into a room with hard surfaces. If space is short, group plants in one container to create a natural, organic look. Couple spiky plants with light, airy foliage plants and large glossy leaves. A mixed display will cast great shadows on interior walls if the lighting is angled appropriately.

Dumb cane's flat leaves contrast with the foliage of the pleated bird's nest fern.

Plants of different shape, size, and leaf texture can combine well together.

Make life easier by placing plants that enjoy high humidity together, such as the tail flower (centre) and air plants, so that they can be misted at the same time.

GROWING AND MAINTAINING

Caring for houseplants is usually easy and almost always great fun. Given the right care and conditions, many of them will become lifelong companions. Each type of plant requires slightly different treatment, with some plants thriving on neglect and others vying for constant attention. Inevitably, they sometimes struggle, but understanding common pests and disease helps to solve problems quickly. Once mature, your plants can be propagated to produce more plants for free. Creating new plants is hugely rewarding and soon becomes second nature.

BUYING YOUR PLANTS

Selecting plants from the vast range on offer can be a daunting task. The choices you make will reflect your style, your budget, and the mood you want to create. However, before heading out to pick up your favourite plants, think about where to buy them, what to look out for, and the best place to put them.

If they're destined for a sunny room, you can discount shade-lovers, and vice versa. Also consider space: do you want to decorate a small windowsill or create a large display in a hallway? Whatever your vision, with some prior thought, buying your plants can be a straightforward and pleasurable experience.

WHERE TO BUY

Houseplants are sold in numerous outlets, including supermarkets and department and DIY stores – all are great places to browse. These non-traditional houseplant outlets – which attract people of all ages to impulse-buy for their homes and workplaces – are encouraging a new generation of indoor gardeners.

However, when buying for the first time, looking for specific plants, or seeking advice, your first port of call should be a specialist nursery or garden centre. Here you'll receive expert guidance and might be able to order a plant of the preferred height and variety. They're also likely to have healthy plants and a good range of composts, feeds, pots, tools, and other accessories.

When choosing a plant, look for healthy specimens with an even shape.

Always make sure that succulents have a good, strong root system before buying them in a shop.

CHOOSING HEALTHY, PEST-FREE PLANTS

Before committing to buying a plant, look at its overall health and the health of the plants it has been placed with in the shop. Check the underside of leaves, where pests most often lurk.

Avoid houseplants that have been displayed by a draughty door – they'll soon fail when you get them home.

For the same reason, in autumn and winter, steer clear of plants that have been displayed outside in the day and moved back indoors at night. Changes in temperature can be fatal.

Houseplants vary in their specific requirements, so a shop display in a south-facing window or a dark corner at the back of the shop will cause damage to some plants. The ideal location for a display of different types of houseplants is in filtered sunlight.

THE JOURNEY HOME

The failure of new houseplants can be due to the journey from the shop rather than the environment at their final destination. When buying a plant in summer, you can simply walk home with it unwrapped. However, if buying in autumn, winter, or spring, take care to protect the plant from the sudden shock of being taken outside and exposed to a considerably lower temperature. Plants such as the poinsettia (see p.85), bought for Christmas cheer, will soon drop their leaves if subjected to even a brief winter chill. Ask the retailer to wrap the plant, or take wrapping paper with you, to keep it well-protected on the journey home.

Request that plants are wrapped by the store in winter.

ONLINE SHOPPING

Shopping for plants online or via mail order is particularly useful if you're keen to source a specific plant from a specialist nursery. Many people aren't able to travel to a distant nursery to purchase one or two plants. An added bonus is that buying plants by post means you can send a houseplant as a gift direct to the recipient's home.

Many online suppliers offer clear images and good descriptions of their houseplants, so it's often a great way for beginners to browse and shop.

If this is your preferred method of buying, make sure you're at home when the plants arrive. Leaving them outside on a doorstep, especially during the

cold winter months, will be detrimental to their health. Also make sure that anyone receiving a plant as a gift is aware a parcel is on its way.

Shopping online is a great way of finding more unusual plants.

NEED TO KNOW

- Choose plants that have a neat and regular shape with new shoots or leaves forming.
- Select plants that are in bud rather than full bloom if you want long-lasting flowers.
- Avoid plants that are either very dry or very wet.
- Check that succulent plants are firmly rooted and make sure they haven't started to rot at the base.

CONTAINERS AND COMPOST

Choosing the right compost and container is key to ensuring a strong, healthy plant that's displayed at its very best. Most houseplants will thrive in multipurpose compost, which is lightweight and clean to use. However, some plants, such as cacti and orchids, need specialist compost for good results. Containers are also vital to the health of a plant and intrinsic to the overall style and mood being created.

Have fun choosing containers and look out for ones that have interesting colours and textures.

PRACTICAL POTS

All plants need good drainage. Soggy, wet compost in a pot with no drainage holes will cause the plant to rot, so unless you're prepared to be very vigilant about watering, grow your plant in a container with drainage holes to allow excess water to drain away. To prevent damaging household goods or causing a slip hazard, place the pot on a drip tray or into a decorative overpot (see *opposite*). You can conceal a pot and drip tray by placing them in a basket to create a jungle look.

When choosing pots, consider their weight, especially if they'll be put on shelving or hung from a hook or beam. They'll also be heavier after watering.

Lightweight pots are the sensible option for hanging or shelf displays.

You can use almost anything as a container, but if it doesn't have drainage holes be careful not to overwater.

GET THE LOOK

When it comes to containers, think outside the box. A trip to a car boot sale or a bric-a-brac or vintage store may result in finding an old household item that could be repurposed as a stylish houseplant container. Look for a pot that will suit your interior and the scale and shape of the plant. However, if you choose to plant directly into a glass or china container with no drainage holes, water with caution as overwatering can be disastrous (see *left*).

Alternatively, opt for specially designed containers such as hanging baskets, terrariums, or bottle gardens (see *pp.18–19*). There's also often a good range of overpots (see *opposite*) in various colours, shapes, and sizes at garden centres and online that will help you to create striking displays.

CHOOSING A COMPOST

Compost is widely available in bags of various sizes. Wherever possible, choose peat-free varieties of the composts below to avoid contributing to the negative environmental impact of peat. It's always handy to keep a small bag of compost in reserve as it doesn't go off or have a use-by date.

COMPOST TYPE	CONTAINS	SUITABLE FOR
Ericaceous (lime-free)	Natural materials that have a high acid content.	Acid-loving/lime-hating plants.
Loam-based	Loam, sand, grit, and natural materials (this is a heavy compost).	Large, top-heavy plants that won't need regular repotting.
Multipurpose	Natural materials such as composted bark. Some of these composts contain slow-release plant feeds.	Most houseplants.
Seed & cuttings	Loam, sand, and natural materials. Fine-textured to encourage even seed-germination. Low nutrient content.	Growing plants from seeds or cuttings.
Specialist (for specific plant groups)	Various materials, according to plant type (usually to improve drainage or retain moisture), including bark, grit, and sand.	Bonsai, cacti, carnivorous plants, citrus plants, and orchids, among others.

Whichever compost you use, always rub out any lumps before planting houseplants.

MAKE YOUR OWN MIX

For the best results, some plants benefit from having additional natural materials in the compost to improve drainage or aeration (see *pp.50–141 for specific requirements*). Rub out any lumps from the compost before using it to repot.

Vermiculite and perlite are often mixed into a compost to improve aeration and retain moisture and nutrients. Choose either one as both these lightweight materials give similar results. Use in a 1:10 mix of vermiculite or perlite and compost.

If you aren't buying cactus compost and want to convert a multipurpose compost to a more suitable medium for succulents, combine coarse sand or grit and perlite into the compost in a 3:2:1 mix of compost, grit or sand, and perlite. Use horticultural-grade sand and grit as builder's sand contains harmful amounts of lime for plants.

> **TOP TIP** PLACE GARDEN-COLLECTED MOSS ON TOP OF COMPOST FOR A DECORATIVE LOOK. IF KEPT MISTED, IT WILL HELP TO RAISE HUMIDITY.

NEED TO KNOW
- If a compost has slow-release feed included, don't give your plants additional feed.
- Store compost in a dry place.
- Keep compost that you've mixed yourself in clearly labelled bags.
- Some orchids, such as the moth orchid (see *p.118*), need to be grown in clear pots as their roots require the light.
- You can place a plant in its plastic pot inside a more decorative outer container (also known as an overpot or sleeve) that doesn't have drainage holes. Always check that your selected overpot isn't porous before putting it on a carpet or a favourite piece of furniture.

Perlite is mixed with composts to improve aeration and drainage.

A layer of garden-collected moss around carnivorous plants helps humidity.

REPOTTING

Over time, healthy plants become pot-bound and require more space to grow. If a pot is packed with roots there's little room for new growth, and it's tricky to water the plant properly. Repotting is usually done in spring. It's a simple process that involves moving the plant to a pot the next size up and refreshing the compost. It's not essential to repot every year, but it does improve plant health and vigour.

Never plant your houseplant deeper in compost than it was in the previous pot.

REPOTTING SMALL PLANTS

Most plants need repotting annually when very young and then every other year thereafter. Repotting offers the perfect opportunity to remove offsets or divide plants (see *pp.40–41*).

YOU WILL NEED Pot-bound plant • Pot with drainage holes, one size larger than your plant • Gloves (optional). • Wipeable surface • Suitable compost • Watering can

HOW TO REPOT

1 Check to see if your plant needs repotting – a sure sign is when the roots are growing out of the holes at the bottom of the pot.

2 Water the plant well a few hours before repotting. If the new pot you've selected has been used before, wash it in warm, soapy water.

3 Rub the lumps out of the compost and add a layer of it to the bottom of the pot (wear gloves if necessary). Carefully remove the plant from its pot and gently tease out the roots. Place the plant in the centre of the pot to check it will sit just below the rim when planted – this will allow room for water.

4 Fill in around the edge of the pot with compost and press firmly in place. The plant must never be planted any deeper than it was in its original pot. Water the repotted plant lightly.

TOP-DRESSING LARGE HOUSEPLANTS

Large houseplants in big containers, or those trained to grow up walls and trellises, are difficult to repot. Instead of increasing the size of the pot year after year, prune the roots lightly so the plant can return to the same pot.

Alternatively, if repotting is tricky and the plant isn't too pot-bound (filling the pot with roots), scrape the loose compost from the top of the pot and replace it with a fresh layer of compost. This can be done very easily every year without disrupting the plant.

If large specimens are planted in a container that's narrower at the top than at the bottom, then adding fresh top-dressing might be the only option as it's very difficult to avoid breaking the container when removing the plants.

Add fresh compost to the top of a pot, but don't fill it higher than it was before.

REPOTTING CACTI

The biggest challenge when repotting a cactus is to avoid getting the tiny spines in your fingers. The easiest way to avoid this is to use a piece of folded newspaper when removing the plant from its pot. An alternative is to wear gardening gloves, but make sure they're a thick leather pair or they won't be spine-proof.

As cacti prefer to be kept on the dry side they come out of their pots easily. Once out of the container, replant the cactus in a pot that's one size larger and filled with cactus compost. Make a hole in the compost and use newspaper to lower the plant into it. Firm in place wearing gloves. Never plant cacti and other succulents too deeply as this can lead to rot.

Protect your hands when repotting cactus by using newspaper to lift the plant.

REUSING OLD COMPOST

Don't throw away any healthy old compost that's left over after repotting your plants. It shouldn't be used again for houseplants, but if you have a garden, spread it on the flower beds or add it to the compost heap to help improve your garden soil. However, compost from diseased or pest-infested plants should be thrown into a bin as soon as possible.

Old compost can be used as mulch in the garden or dug in to improve your soil.

> **TOP TIP** NEVER FILL A CONTAINER RIGHT TO THE RIM WITH COMPOST. LEAVE ABOUT 2CM (¾IN) AT THE TOP TO ALLOW FOR WATERING, OTHERWISE THE WATER WILL SPILL OUT OVER THE SIDES.

NEED TO KNOW
- When removing pot-bound plants from flexible plastic pots, squeeze the sides of the pot to loosen the rootball and, if possible, push the plant out from the bottom rather than tugging on the foliage.
- If a plant's roots are growing out of the drainage holes of its container, it's a sure sign that repotting is due.
- When water is slow to soak into the compost, it can be an indication that the container is completely packed with plant roots.
- Yellowing or wilting leaves are often a warning that a plant is pot-bound.
- Some plants react well to being slightly pot-bound as it prompts them to flower (see plant profiles for guidance, *pp.50–141*).

TOOLS AND EQUIPMENT

Having the right tools and equipment to hand will make caring for your houseplants easier and allow you to take up propagation opportunities without any prior planning. Not all indoor gardeners will require every tool suggested here – your handy tool kit can be tailored to meet your specific needs and won't take up too much room in the house. Good-quality, well-made tools are a great investment and should last for years.

Children enjoy using mini trowels and forks when planting and repotting.

CUTTING TOOLS

The most important cutting tool in a garden is a pair of secateurs. These are also very handy indoors if you're growing larger specimen plants that need hard pruning. However, most indoor gardeners will find that a pair of florist or gardening scissors is far more useful. Scissors can be used to remove damaged leaves, trim off brown leaf tips, and deadhead flowers.

A penknife is also an invaluable cutting tool. Plants such as begonias (see p.59) and African violets (see p.126) can be propagated by leaf cuttings (see p.38) and a penknife is essential for this task.

Buying good-quality cutting tools and caring for them will prevent you from having to get replacements. After every use, carefully wipe the blade of the tool with an oily rag to keep it sharp and clean. Store your clean tools safely and well out of the reach of children.

TOP TIP KEEP YOUR TOOLS TOGETHER IN A BUCKET SO THEY'RE EASY TO CARRY AROUND THE HOUSE.

ESSENTIAL EQUIPMENT

Bottle-garden tools • Cloths to clean leaves and tools • Expanded clay granules • Gloves • Green cane (with cane topper) for supporting orchids • Jam jar for cuttings • Labels • Mister for raising humidity • Moss pole for supporting climbers • Paintbrush for cleaning cacti • Penknife for leaf cuttings • Plant ties for climbers • Pots and trays for propagation • Scissors for deadheading • Secateurs for pruning • Thermometer for room temperature • Trowel for planting • Watering can

A sharp, clean pair of secateurs is a useful tool for indoor gardeners.

A sharp penknife is essential for taking houseplant cuttings.

PROPAGATION KIT

There's a huge sense of achievement when you successfully take cuttings, divide plants, or sow seeds. Having the right tools will greatly increase your chances of success in these tasks. The most important thing is to make sure your equipment is spotless in order to prevent spreading plant pests and diseases to vulnerable cuttings or seedlings. Always wash your pots and trays well and wipe down surfaces and cutting tools before starting work.

To avoid damaging your kitchen worktops or tables, invest in a large watertight tray on which to divide plants and take cuttings. Another essential item is a watering can with a fine rose attached, especially if you're growing from seed as a fine spray of water is less likely to dislodge or disturb seedlings.

Other useful propagation tools are scissors, a penknife, labels to date and name young plants, and a thermometer to ensure you're providing the right temperature. Old jam jars, filled with water, can be used to strike cuttings.

Use a pair of sharp scissors to cut soft stems for propagation.

WATERING EQUIPMENT

Make life easy for yourself and buy a lightweight watering can for your indoor garden. A can with a top and side handle will make watering and filling up easier. Make sure your can will fit under all the taps in the house to prevent you from having to go up and down stairs with cans of water.

A long spout is ideal as this makes it far simpler to reach containers that are positioned high up or at the back of a display. Choose a can with a detachable rose for watering seedlings; this attachment isn't necessary when watering mature plants.

A detachable rose allows gardeners to water their seedlings with a fine spray.

Small trowels and forks are very useful when planting bottle gardens.

SPECIALIST TOOLS

Gardeners who want to create a bottle garden should invest in one or two specialist tools. These are basically mini trowels and forks with long handles. If your hands won't fit in the bottle, you'll need them to plant up. Another option is to make your own tools by strapping a short cane to the handle of an old kitchen fork or wooden spoon with tape – this will extend the reach.

A paintbrush is a vital tool if you're growing cacti. It's almost impossible to repot these plants without getting compost stuck in their spines. A paintbrush is the only sensible and practical way of cleaning them up.

CHOOSING THE RIGHT GLOVES

Some houseplants are toxic, prickly, or can cause irritation to the skin, so good gardening gloves are vital to keep your hands safe and clean. Choose a pair that fits well and allows you to feel what you're doing. Thick leather gloves are useful for handling cacti. Bamboo gloves allow great dexterity and are made of a sustainable material. Gloves with a rippled latex palm will give you a good grip when lifting heavy pots. Always check that the gloves you intend to buy are washable.

Protect your hands by wearing a comfortable pair of gardening gloves.

WATERING AND FEEDING

The trick to successful watering is to understand the natural habitat of your houseplant and try to mimic it. Over- and underwatering are the most likely causes of failure, so be aware of your plants' requirements, while incorrect feeding can also cause harm. Some plants require specialist feeds – again, be conscious of their needs. Follow a few simple rules and you'll be rewarded with strong, healthy plants.

Watering cans with a long spout are ideal for hanging baskets. Houseplants depend on us for all their needs.

WHEN TO WATER

Every type of houseplant is different: foliage plants soon wilt if they need watering, whereas succulents don't display such obvious signs. It's risky to allow plants to reach wilting point as this will have a long-term impact on their health; if badly wilted, they won't revive. Plants from a tropical rainforest are likely to need greater humidity and more regular watering than those originating from a desert habitat.

Test whether your plants need watering by pushing your finger about 5cm (2in) into the compost. If it's moist, hold off on watering (check individual plant profiles for advice: see pp.50–141). Lifting pots up to feel the weight of a plant is also a good watering indicator. Very light pots mean dry compost.

Push your finger into the compost to establish if the plant needs watering.

Never let your houseplant sit in water – always pour away excess.

HOW TO WATER

All indoor gardeners need a small watering can with a long spout – this will allow you to direct the water to the correct place. Get into the habit of refilling the can and leaving it on a warm windowsill as houseplants prefer tepid water – but be aware that some plants won't tolerate tap water (see opposite).

If houseplants have become too dry, they may require double watering. To do this, place the plant on a draining board, water well, allow the water to

Place plants on a draining board or tray when watering so that excess can drain off.

drain, then water and drain again. Never leave your plants standing in a saucer of water – their roots must have air to function properly.

Some plants such as bromeliads (see p.111) require water to be poured into their central crown; others, including African violets (see p.126), will show significant damage on their leaves if their foliage is watered. Succulents are often grown in cactus compost as this encourages good drainage and therefore goes some way in protecting the plant from overwatering.

HOLIDAY WATERING

Don't leave your houseplants without water while you're away from home. Either ask a trusted neighbour to care for them or invest in watering spikes: these are filled with water and release it gradually into the compost. Some gardeners push one end of a thick cotton cord into the compost and the other end into a bottle of water. The cord acts as a wick and stops the compost from drying out.

Before going away, move your plants out of very bright sunlight and give them a good watering.

NEED TO KNOW

Some plants, such as ferns, aren't damaged by having wet foliage and can be watered from above. Others must be watered from below or their leaves will suffer: place a saucer of water under the plant pot (it must have drainage holes), leave for about 15 minutes, then discard any excess.

If you're going away during the summer, move your plants clear of south-facing windows to reduce their need for water.

TOP TIP NOT ALL PLANTS CAN BE WATERED WITH TAP WATER: SOME, SUCH AS CARNIVOROUS PLANTS, NEED RAINWATER, WHILE OTHERS REQUIRE DISTILLED WATER.

FEEDING PLANTS

Most houseplants require feeding with a balanced liquid fertilizer in spring and summer (the typical growing season). Feed plants that are grown purely for their flowers with a high-potash feed to encourage flowering. Orchids, citrus, and cacti must be fed with specialist feeds. Always follow the instructions on the packaging carefully, and never overfeed plants, as this can damage them – if in doubt, err on the side of caution by slightly underfeeding.

For easy application, choose a liquid concentrated feed that can be diluted in water and watered on your plants with a can (see *image, left*).

Fertilizer contains three main nutrients – nitrogen, phosphorous, and potassium – along with some of the trace elements that plants need. Plant food packaging shows the ratio of these main nutrients: a balanced or general houseplant feed has roughly equal amounts of all three nutrients, whereas specialist feeds contain a dominant ingredient (see *below*).

A diluted, liquid concentrated feed is easy to apply with a watering can.

TYPE OF FEED	DOMINANT NUTRIENT	PROMOTES
Balanced feed	N, P, K	Healthy, plentiful foliage
Cactus feed	P	Strong, healthy roots
Citrus feed	K	Flowers and fruit

KEY Nitrogen (N), Phosphorous (P), Potassium (K)

PRUNING AND CLEANING

It's important to keep houseplants clean and neatly shaped – and not just for their decorative appeal. Thick layers of dust on leaves can hinder photosynthesis and result in poor plant health. Pruning and cleaning your houseplants will also encourage strong, new growth, and ensures that large specimens can be kept to the required size and remain impressive for years to come.

When pruning, take a step back from the plant from time to time to make sure the overall shape is looking good.

PRUNING YOUR PLANTS

The aim of pruning is to ensure that the plant has a neat shape and is left with only strong and healthy stems. Removing untidy growth keeps houseplants compact and healthy and also encourages them to be bushier. Pruning off dead and damaged leaves helps to prevent the spread of pests and other plant problems, while removing faded flower stems prompts more blooms to grow.

When pruning your plants, always use sharp, clean secateurs for woody growth and scissors for cutting off softer leaves (see pp.32–33). Make your pruning cuts just above a leaf node (a bump in the stem where new growth forms). When cutting off an entire stem, cut as close as possible to the main stem as no plant looks good covered in little stumps. If removing a leaf from a plant that grows from the base, cut off the leaf at soil level to keep the plant looking tidy.

Wear gloves and use sharp, clean secateurs for pruning the plants.

CARE WHEN PRUNING

Before reaching for the secateurs, look at the plant carefully from all angles and turn the pot around. Try to get a clear idea of what it will look like once you've removed certain branches, leaves, or stems, and make sure the overall appearance will be improved by the cuts you intend to make. After each cut, step back and review the plant again.

Always use a pair of sharp scissors or secateurs. Blunt cutting tools will leave stems crushed or torn, making the plant untidy and also more susceptible to pests and diseases.

Wearing gloves is a sensible precaution when pruning as the sap of some plants will irritate the skin.

Make sure you give the plant a neat and natural shape when pruning.

Deadheading will keep plants tidy and encourage a second flush of flowers.

REJUVENATING
HOUSEPLANTS

You can give a new lease of life to some houseplants by pruning them drastically. This is often done as a last attempt to keep a plant a bit longer before composting it or because it looks unhealthy or is very leggy.

Cut back all the stems hard (to just above soil level) to see if it prompts fresh growth. You can prune indoor plants at any time of the year, but if pruning to rejuvenate them, choose late winter or early spring. Plants that respond well to this treatment include the Madagascar dragon plant (see p.80) and the umbrella tree (see p.130).

Umbrella trees respond extremely well to a hard prune.

NEED TO KNOW

- Remove faded flower stems to encourage reflowering.
- Cut out stems on variegated plants that have reverted to plain green.
- Prune stems that are causing the plant to look unbalanced and untidy.
- Cut back lead stems to keep plants to a manageable size.
- Prune or pull off leaves that are yellow and unhealthy.
- Cut off brown leaf tips.

TOP TIP AVOID SPREADING PESTS AND DISEASES FROM ONE PLANT TO ANOTHER BY CLEANING SECATEURS AND SCISSORS WITH A HOUSEHOLD DISINFECTANT AFTER USE AND THEN DRYING THEM OFF. SHARPENING TOOLS ARE AVAILABLE FROM GARDEN CENTRES AND OTHER OUTLETS.

CLEANING FOLIAGE

Houseplants are dust magnets, but foliage can be refreshed by a regular clean with a soft, damp cloth – a quick wipe should leave it looking at its best. A thick layer of dust stops the leaf absorbing sunlight and reduces the plant's ability to perform photosynthesis (the process by which plants convert energy from sunlight into food for growth). Clean the underside of the leaves too, as this is where pests lurk.

Avoid wiping the leaves of plants that have fluffy foliage, such as African violets (see p.126), as this will damage them. Use a paintbrush to remove dust and compost from spiky cacti – it's the perfect tool for this task.

It's important to wash your cleaning cloths regularly to prevent the spread of pests and diseases from one plant to another. If you have several plants, you'll need more than one cloth.

(*above*) **Remove dust or compost** from cacti using a paintbrush.

(*left*) **Wipe foliage** over gently with a soft, damp cloth to get rid of any dust.

PROPAGATING BY LEAF AND STEM CUTTINGS

Increasing your houseplant collection by propagation is immensely rewarding. Different plants are suited to different propagation methods: soft-stemmed plants give great results from stem cuttings, while some plants with fleshy leaves respond well to leaf cuttings. For best results, take cuttings in spring or early summer from healthy plants and expect to see new growth in six to eight weeks.

Propagating your own plants is great fun and also saves money.

BEGONIA LEAF CUTTINGS

Begonias have generous leaves, so more than one plant can be created out of just one leaf using the method below.

YOU WILL NEED Penknife • Clean pot or seed tray • Cutting compost with a handful of perlite • Watering can with a fine rose • Clear plastic bag

HOW TO REPOT

1 Water the plant an hour before taking cuttings. Remove a healthy leaf by using a penknife to make a clean cut at the base of a stem.
2 Cut out the stem and centre of the leaf. Divide the leaf into three or four sections measuring about 2.5cm (1in) in length and width. All sections must have a vein running through them.
3 Fill a seed tray or pot with the cutting compost and perlite mix. Firm down, then push each leaf section into the compost just enough so they can stand up. The veins that were cut through must have contact with the compost.
4 Water using a fine rose. Cover the pot with a clear plastic bag and keep in a warm room away from direct sunlight. When two to three new leaves appear, the cuttings can be potted on.

OTHER LEAF-CUTTING TECHNIQUES

The leaves of the Cape primrose (see p.136) are long with a strong central vein. Cut the leaf into three to six sections. Insert the leaf pieces into a mix of cutting compost and perlite just enough so they stand up, making sure that the lower part of the leaf is in the compost. Cover with a clear plastic bag until new leaves form.

The variegated snake plant (see p.128) can be treated in a similar way. Cut a healthy leaf horizontally into chunks of 5cm (2in). Then position the sections so that the lower part of each one is pushed into the compost.

When taking leaf cuttings of succulents, remove the leaf as a whole. Leave it to dry for 24 hours. Fill a pot with cactus compost mixed with a handful of horticultural sand and then insert the bottom of the leaf into the compost so that it stands up. Place in a sunny spot. Don't cover the pot with a plastic bag.

Plant Cape primrose leaf sections with the lower part in the compost.

Remove the lower leaves of each cutting to leave a length of clean stem.

Once cuttings have been inserted in the container, water using a rose on the can.

STEM CUTTINGS

Young, healthy growth produced in spring or summer (from *Tradescantia*, for example, *see above*) is ideal for stem cuttings. Select a stem that's not flowering and remove a length of about 12cm (5in) by cutting just above a leaf joint on the parent plant. Remove the lower leaves and trim each cutting to just below a node. Some gardeners dip the end of the stem cutting into hormone rooting powder to speed up root growth, but it isn't essential.

Place the cuttings into a pot of cutting compost, water lightly, and cover with a clear plastic bag. Leave in a position of filtered sunlight.

Drop your cuttings into a glass container of water and watch the roots growing.

ROOTING CUTTINGS IN WATER

The cuttings of many houseplants respond well to being rooted in water rather than compost. The benefit of this technique is that you can enjoy watching the roots gradually forming and keep a close eye on progress. Children in particular love propagating plants in this way.

Take a stem cutting as you would if it were to be grown in compost, but drop it into a jar of water. Replace the water every few days and when the cutting has a healthy root system, move it to a container of compost to grow on.

RAISING FROM SEEDS, OFFSETS, AND DIVISION

The easiest way to produce more plants is to grow offsets on and divide mature specimens. Some plants that die after flowering, such as bromeliads, grow offshoots at their base, while others, like spider plants, hold them on stems. Raising from seed requires patience, but it's a great way of producing lots of plants on a tight budget. Whichever way you choose, it's vital your tools and pots are clean.

Propagating houseplants is a highly addictive but healthy hobby.

REMOVING OFFSETS

An offset is a baby plant. Offets that are produced at the base of the plant can be cut off when plants are repotted in spring and used for propagation. A good example is the bromeliad.

YOU WILL NEED Clean penknife • Hormone rooting powder (optional) • Cutting compost with a handful of perlite mixed in • Container • Short garden cane • String • Watering can

HOW TO GROW PLANTS FROM OFFSETS

1 Wait until the offsets are one-third the size of the parent plant. Carefully remove the parent plant from its pot.

2 Remove loose compost from the rootball. Use a penknife to cut off the offsets as close as possible to the parent plant.

3 Dust the end of the offset with hormone rooting powder (optional). Push the offset into a container of pre-watered cutting compost. It's normal for offsets to have no roots at this stage. Don't push the plant too deeply into the compost as it will rot.

4 Offer the plant the support of a cane if it topples and place in a room with filtered sunlight. Expect to see signs of growth in approximately six weeks.

To speed up root growth, offsets are left attached to the parent plant.

POTTING UP HANGING OFFSETS

Plants that produce plantlets (hanging offsets) held on the end of long stems are extremely easy to propagate from. Spider plants (see p.65) and creeping saxifrage (see p.129) are among the plants that propagate in this way. Choose an offset with healthy foliage and signs of roots growing at the base. Fill a pot with cutting compost and push the offset into it, although not too deeply. Make sure it's still firmly attached to the parent. Leave it attached until you see signs of healthy new growth and then cut it free.

DIVIDING PLANTS

The quickest way to produce new plants is to divide mature specimens. Plants that have fibrous roots and produce new stems readily such as the rattlesnake plant (see p.93) are good candidates.

Water the plant well and leave for an hour, then gently slide the plant out of its container and remove any loose compost. Carefully tease the plant roots apart to make two or three separate plants. Make sure each one has a good portion of the roots. If the roots are very tightly packed, you may need a sharp knife to cut through them. Repot each plant separately, ensuring they're potted at the same level as in their previous pot. Water well and place in a room that has filtered sunlight.

Spend time preparing an even seed bed before sowing.

When moving seedlings to modules, hold them by the leaf, never the stem.

GROWING FROM SEED

Growing plants from seed is the quickest, easiest, and most cost-effective way of propagating a large quantity of houseplants that are grown as annuals, such as coleus (see p.132).

The technique may differ slightly for each type of plant, but the basics remain the same. Fill a pot or seed tray with seed compost; level it out, firm it down slightly, and water with a can that has a rose. Place the seeds evenly on the compost. Cover them with a thin layer of compost (depth will vary according to seed). Place a plastic hood or clear plastic over the top and keep at the recommended temperature in a room that has filtered sunlight.

Once the first leaves emerge, move seedlings to individual pots or modules. Make holes in the compost for their roots, then plant the seedlings in it. Place in a room with filtered sun, where they'll develop into mature plants.

Tease apart the parent plant, making sure each part has a good section of root.

For success, only divide mature plants that have a generous root system.

UNUSUAL GROWING TECHNIQUES

More people than ever are turning to indoor gardening, and gardeners themselves are becoming increasingly adventurous in how they grow and display their plants. Architects and interior designers are routinely including indoor planting schemes in their blueprints for buildings, and the range of containers, houseplant lighting, and supports is expanding rapidly to meet demand.

Modern homes are now often designed to allow maximum light into rooms, allowing houseplants to thrive.

If growing hyacinth bulbs in water, position them just above the surface.

PLANTING IN WATER

Hydroponics is the term used to describe growing plants in water. Not only is it one of the easiest methods of growing some houseplants, it also has immense visual impact: instead of hiding plant roots in a container of compost, you can make them visible by using a see-through pot (often glass). Decorative pebbles to help support the plants in the water add to the effect. Use water-soluble fertilizer in spring and summer and replace the water regularly to keep it clean.

GROW LIGHTS

Domestic lighting is sometimes used to create greater impact in a room and dramatize plants – but not all household lighting (domestic light bulbs, for example) will assist plant growth, and some can even scorch the leaves if they're too close. Specialist lighting systems are, however, widely available.

These "grow lights" are designed to provide the right type and quality of lights for plants and encourage sun-loving plants to grow in areas where natural light is limited. They're also used to light up displays without increasing the temperature or risking foliage scorch. Keep a look out for plant-boosting lights that will also contribute to the ambience of a room.

Dramatize your plant displays at night by using artful lighting techniques.

Invest in horticultural lights if growing plants in a shady room.

LIVING WALLS

Indoor gardeners often have to try to fit as many plants as possible into cramped spaces. A living wall is a great solution to this problem. Specially designed wall-hanging planters are available for this style of display, including ones with individual modules that allow you to easily add and replace plants as needed. However, it's vital you choose the right plants for the right wall (see pp.14–17).

Combine plants that need the same care and growing environment and give them space – tightly packed displays won't remain healthy.

Air plants that grow happily without soil (see p.138) can be used to cover a wall if you tie them into a frame with string. This is an easy display option as the frame can be removed to both mist and water the plants without causing damage to plasterwork.

Tie air plants to a frame with string and display it on a wall.

INSIDE-OUT PLANTS

Gardeners with a small outdoor space soon realize that some houseplants can easily be moved outside during the summer months. Use heavier pots if you're considering doing this to prevent the plants from being damaged by winds. You can use lighter pots if you keep an eye on the weather and move them back inside if heavy rain or wind are forecast. Light rain and fresh air are good for plants and keep them clean.

TOP TIP MAKE MOVING HEAVY PLANTS OUTSIDE EASY BY PLACING THEM ON POT TROLLEYS. WHEN LIFTING WEIGHTY POTS, ASK A FRIEND TO HELP, AND ALWAYS BEND YOUR KNEES WHEN LIFTING.

Houseplants will often thrive outside in summer and may reward you with a spectacular, tropical display.

HOUSEPLANT PROBLEMS

Houseplants are generally easy to grow but will sometimes suffer due to care and environmental issues, pests, and diseases. The common indicators of an ailing plant are leaf drop, unhealthy-looking foliage, lack of flowers, and spindly growth.

Once you've eliminated the possibility that pests or diseases are not the cause of an issue, look carefully at watering, feeding, light levels, draughts, and sudden changes in temperature. Making simple adjustments can often greatly improve a plant's health.

AVOIDING TROUBLE

The conditions in our homes are very different from the native habitat of most houseplants. It's therefore no surprise that some plants will suffer if the room environment is alien to them. If you spot a problem, act quickly to identify a possible cause by researching the plant's requirements (see pp.50–141).

It's quite common for plants to suddenly show signs of stress if the temperature dramatically changes in a room – for example, when you turn on the heating in autumn or light a fire in winter. Before changing the room temperatures or light levels, think about the possible effect on your plants and, if necessary, move them elsewhere.

Houseplants also react badly to being knocked or brushed past regularly – avoid damaging their leaves by making sure they have sufficient space.

Protect plants from damage and extreme changes in temperature when transporting them.

NEED TO KNOW
- If a plant has been kept in a draughty shop it's common for it to look unwell until it settles into your home.
- Discard leaves or flowers that have dropped onto the plant compost as these can rot and cause problems.
- Some plants, such as bromeliads (see p.111), die after flowering, so their demise may be unrelated to care or the room environment.

There are many possible reasons for yellowing leaves, so discovering the exact cause of the problem may take some time.

COMMON PROBLEMS

Issues with leaves, flowers, and stems are clear indicators that a plant is ailing and prompt action should be taken.

FLOWER PROBLEMS

NO FLOWERS In order to flower, some plants need to be pot-bound or to have reached a certain maturity. Lack of flowers can also be due to poor light levels, not enough hours of light, dry air, or overfeeding. If buds form and then drop, this may be caused by dry air, underwatering, a shortage of light, or moving to a different environment.

GROWTH PROBLEMS

SPINDLY GROWTH This is often an issue in winter and early spring as plants may have suffered from low light levels in winter. In their search for light, they grow tall and thin. Overwatering might have played its part – be vigilant about not overwatering in winter. In spring, repot, prune out any very spindly growth, and start to feed your plant throughout the spring and summer.

LEAF AND STEM PROBLEMS

BROWN LEAVES Leaf tips can turn brown if they're often knocked by passers-by or if the air in the room is too dry. Misting plants will resolve the latter. Cut off brown tips with scissors. If a leaf turns completely brown it may be due to underwatering, too little light, or too much heat. Resolve the issue by discovering the cause and adjusting conditions accordingly.

SUDDEN LEAF DROP This is the result of shock. Plants dislike sudden changes in light and temperature and will be affected by cold draughts. Gardeners may be tempted to increase watering and feeding when leaves start to drop, but be cautious of this as overwatering and feeding can lead to further leaf drop. Check your plant has the right environment and wait for recovery.

WILTING LEAVES AND STEMS The likely cause of this is too much or too little water. If a plant has been left for too long without water, re-wet the compost quickly by submerging the pot in a bucket of water. When air bubbles stop rising to the surface, the compost is completely wet. Lift the plant out of the bucket and leave to drain.

YELLOW LEAVES There are many reasons for this, including overfeeding, over- or underwatering, cold draughts, and red spider mites (see p.47). All plants lose leaves as part of their natural life cycle, so the odd yellow leaf may not be something to worry about. Lower leaves are often shed, especially on trunk-forming plants such as the Madagascar dragon plant (see p.80).

PESTS AND DISEASES

Most plants are affected by pests and diseases at some point. The sooner you spot problems, the easier it is to resolve them. If you leave them to develop, you may have to isolate or discard the plant. Many gardeners opt for organic or natural solutions, such as spraying or wiping leaves with soapy water, before resorting to insecticides and fungicides.

GREY MOULD

PROBLEM Fuzzy, grey mould appears on the stems and leaves of affected plants and leads to decay.
CAUSE Humid environments, poor ventilation, overcrowded plants.
REMEDY Remove affected leaves quickly; increase ventilation; don't overcrowd plants. Avoid by deadheading flowers and clearing up fallen leaves regularly.

APHIDS

PROBLEM Stunted and distorted plant growth, spread of viruses, presence of sticky honeydew excrement.
CAUSE Green, black, or brown sap-sucking aphids on plants stressed by their environment (see p.44).
REMEDY Wipe off if caught early, or spray plants with soap and water solution, or treat with insecticide.

CROWN ROT

PROBLEM Stems weaken and turn brown at the base of the plant. The rot spreads to the leaves and causes decay.
CAUSE Overwatering or using a pot with no drainage holes so that plants become waterlogged.
REMEDY Remove damaged stems and leaves, reduce watering, and change the pot to allow for good drainage.

MEALYBUGS

PROBLEM Stunted and distorted growth; sticky honeydew coating the plant.
CAUSE Sap-sucking insects that resemble fluffy cotton appear on leaf axils and under leaves; some feed on roots.
REMEDY Use a cotton bud to remove. Throw out badly infested plants. Mealybugs are often present on new plants, so check before buying.

POWDERY MILDEW

PROBLEM A white powdery fungus appears on flowers, stems, and leaves. It's unpleasant to look at but not fatal.
CAUSE Lack of water, too much feed, or poor ventilation.
REMEDY Remove affected parts of the plant, water more regularly, and improve ventilation in the room. Avoid getting foliage wet when watering.

SCALE INSECTS

PROBLEM Bumps on stems or under leaves; foliage may turn yellow and be covered in sticky honeydew excrement.
CAUSE Brown, sap-sucking, immobile pests on the underside of leaves.
REMEDY Prune out badly affected stems or dab them with methylated spirits on a cotton bud. The pests are often found on new plants, so check before buying.

VINE WEEVILS

PROBLEM Begonias and cyclamen, among other plants, develop holes on their leaf margins in spring and summer.
CAUSE Adult vine weevils, which are often found on new plants. Grubs eat roots in autumn and winter.
REMEDY Check pots for grubs and under pots for adults. Wash off roots and repot, or throw out the plants.

RED SPIDER MITES

PROBLEM Yellow marks and unsightly webbing appear on foliage.
CAUSE Hot, dry, and overcrowded growing environments attract tiny, fast-spreading red spider mites.
REMEDY Mist plants as the pests prefer a dry atmosphere. Throw away badly affected plants immediately to prevent the spread of these pests.

SOOTY MOULD

PROBLEM Black fungus covers leaves, forming a barrier between the leaf and sunlight and resulting in weak growth.
CAUSE The fungus grows on the sticky honeydew excreted by aphids, scale insects, and mealybugs.
REMEDY Wipe off the fungus with a cloth soaked in warm water. Control the offending insects (*see left and above*).

VIRUSES

PROBLEM Symptoms vary from streaked and spotted leaves to mosaic patterns; flowers can develop white markings. Growth becomes stunted or distorted.
CAUSE Viruses can be brought in by insects or the plant might have been affected before purchase.
REMEDY There is no cure. Discard the plant if you're sure it's a virus.

HOUSEPLANTS A–Z

Trying to decide which houseplant will suit a room can be baffling. With so many indoor plants available at garden centres and nurseries, there's a plant to suit every growing environment and every taste. Our A–Z guide will quickly and easily help indoor gardeners find the right plant for the right place. It features a wide selection of houseplants suitable for sun and shade, as well as plants that trail, climb, or offer spectacular scented blooms. Each plant entry includes invaluable growing and displaying advice.

MAIDENHAIR FERN

ADIANTUM RADDIANUM

The maidenhair fern adds a light and frothy touch to any display, thanks to its delicate foliage. Small, bright green, rounded fronds are held on darker, arching stems. Plants can grow to a significant size and will flourish in any high-humidity environment, such as a large terrarium.

HEIGHT 50cm (20in)
SPREAD 60cm (24in)
FLOWERS None
FOLIAGE Light and airy
LIGHT Filtered sun/light shade
TEMPERATURE 10–24°C (50–75°F)
CARE Fairly easy
PLACE OF ORIGIN The Caribbean

CARE

This airy maidenhair fern is ideal for a humid bathroom or kitchen but will also thrive in a warm – but not hot – room, as long as you give it a daily misting with water. It needs plentiful light, but avoid spots that receive direct sun, which will damage foliage.

Plant the fern in multipurpose compost, and be sure to keep this just moist but never wet. Reduce watering slightly in winter if the plant is kept in a cool room. Maidenhair ferns quickly suffer if left to dry out, but in other aspects are easy to maintain, requiring no pruning other than the removal of untidy foliage. Feed the fern monthly with half-strength liquid fertilizer.

If necessary, repot your fern in spring. This is also the time to propagate the plant, if you wish, by carefully dividing the rootball into two or three pieces.

PROBLEM SOLVING Browning foliage is often caused by excess heat, sunlight, or air that is too dry. To remedy this, prune out the dead fronds and move the plant to a cooler, shadier spot. Brown dots on the back of leaves are healthy spores and not pests.

DISPLAY

This fern is perfect for a humid terrarium, where it partners well with the mosaic plant (see p.91) and the radiator plant (see p.117), which offer wonderfully contrasting foliage. The maidenhair fern also looks great in a hanging basket or tabletop pot. Be sure to turn the pot occasionally to promote even growth.

Maidenhair ferns grow fast, so start with a small, quite inexpensive plant.

ALSO TRY

The pretty, divided foliage of the maidenhair fern is great for breaking up and softening the stark outlines of walls, cabinets, and other interior features. Plants with similar attributes include:

- **Sensitive plant** (*Mimosa pudica*), height 70cm (28in). This plant is a talking point for its feathery foliage, which folds in on itself if touched.
- **Tree maidenhair fern** (*Didymochlaena truncatula*), height 75cm (30in). The fronds of this plant are more leathery than those of the maidenhair fern, but are an attractive bronze when young.

The leaves of the sensitive plant fold up at night and when touched.

URN PLANT AECHMEA FASCIATA

This stylish bromeliad has silvery green foliage that forms a well-shaped rosette, or urn, in which it naturally traps water. It's prized mainly for its flower spike of pink bracts and purple flowers, but even without this, the plant makes a big statement in any room.

HEIGHT 60cm (24in)
SPREAD 60cm (24in)
FLOWERS Pink and purple
FOLIAGE Silvery-green
LIGHT Light shade
TEMPERATURE 15–27°C (59–81°F)
CARE Fairly easy
PLACE OF ORIGIN Brazil

Flowers of the urn plant are encouraged by high temperatures.

CARE

In Brazil, this plant grows high on tree trunks, trapping rainwater in its rosette of leaves. Its tiny root system means that it is content in a small pot, making it good for tight spaces. Grow it in an equal mix of orchid and multipurpose compost, feeding in spring and summer. Place it in light shade, away from direct heat or draughts, and water the compost and the rosette of the plant with rainwater, not tap water. Allow the compost to dry out between waterings.

Urn plants are often sold when they're in flower; if there's no flower, be aware that it can take several years to initiate. Flowers are encouraged by high temperatures and, once formed,

may last for months; cut them off when they turn brown and fade. After flowering, the plant will start to slowly die but don't fear – after a couple of months, offshoots will form around the base. These can easily be separated from the parent and potted on.

PROBLEM SOLVING The urn plant is fairly problem-free. Pour out any stale water from the crown and replace with fresh; avoid overwatering as this can cause plants to rot. If leaves develop brown tips, dry air is the likely cause; mist the plants to avoid this issue.

DISPLAY

Bromeliads look great in individual pots; a cluster of them will give your home a distinctly tropical look. Try growing the

Display urn plants in sturdy containers as they can become quite top-heavy.

Plants will need repotting once every two or three years.

urn plant alongside scarlet star (see p.94). Its leathery, green leaves will contrast well with the distinctive pastel green and silver colours of the urn plant.

ALSO TRY

If you've fallen for the charms of the urn plant's silvery-green foliage, try these specimens, which have similarly metallic-looking leaves:
- **Aluminium plant** (*Pilea cadierei*), height 30cm (12in). This low-growing trailing plant is sturdy and great for beginners. Like the urn plant, it enjoys a spot out of direct sunlight.
- **Spider aloe** (*Aloe humilis*), height 30cm (12in). This neat little houseplant is perfect for a tight but sunny space.

CHINESE EVERGREEN

AGLAONEMA COMMUTATUM

This durable houseplant is almost foolproof: it tolerates a wide range of temperatures and humidities, and survives in low-light conditions where many other plants fail. It's grown for its attractive leathery, spear-shaped, variegated silver and green foliage, and its beautiful arching habit.

HEIGHT 45cm (18in)
SPREAD 45cm (18in)
FLOWERS Unlikely to bloom indoors
FOLIAGE Variegated
LIGHT Shade
TEMPERATURE 16–25°C (61–77°F)
CARE Easy
PLACE OF ORIGIN The Philippines; widely cultivated in China

A tray filled with water and pebbles will prevent your plant sitting in water.

CARE

The native habitat of this plant is the shady and moist understorey of tropical woodlands. When growing at home, try to mimic these conditions, keeping the air humid by misting the plant with a fine spray twice a week.

While the Chinese evergreen copes well in dim conditions, it won't survive without any natural light. A bathroom that has a skylight or a frosted window would make a perfect home. Keep it out of direct sunlight and well away from draughts, and also ensure the compost is moist all year round, but never leave this plant sitting in water. Plant in multipurpose compost mixed with a handful of perlite (see p.29), which will help to keep moisture and air around the roots. Feed once a year during the springtime using normal, soluble, houseplant fertilizer.

The Chinese evergreen is a relatively slow-growing plant; to allow it to reach its mature size, it needs repotting only every other year.

PROBLEM SOLVING If the air in the room is too dry, cool, or draughty, foliage can become brown at the edges. Either remove the affected leaf or use scissors to snip off the browning tips, leaving the leaf with a pointed end. Mealybugs can also be a problem with this plant; they're often found at the base of stems but are easily spotted (see p.46).

DISPLAY

There are few plants, other than the cast iron plant (see p.56), that can be grown alongside the Chinese evergreen in shade. They work well together, with the plain green cast iron plant offering a wonderful backdrop to the silvery foliage of the Chinese evergreen. Older plants may flower, producing blooms similar to those of the peace lily (see p.133), but these rarely appear – the main attraction is the striking foliage.

ALSO TRY

Spear-shaped leaves always offer a statement and structure to a room setting. If you're looking for other plants that offer this shape of leaf, attractive variegations, and also thrive in light shade, then try:

- **Amazonian elephant's ear** (*Alocasia × amazonica*), height 1.2m (4ft). This dramatic plant has large, very shiny, dark green leaves with bold, pale green veins.

Amazonian elephant's ear is a stunning plant with dramatic foliage.

ALOE VERA *ALOE VERA*

Prized by some for its healing abilities, this succulent is among the most commonly bought houseplants. In the wild, aloe vera can grow up to 1m (3ft 3in) wide; indoors, its size is generally more compact, but its long, fleshy leaves still catch the eye.

HEIGHT	60cm (24in)
SPREAD	60cm (24in)
FLOWERS	Spikes of yellow/orange
FOLIAGE	Succulent
LIGHT	Filtered sun
TEMPERATURE	10–27°C (50–81°F)
CARE	Easy
PLACE OF ORIGIN	Africa and the Arabian Peninsula

Aloes are often kept in the kitchen so that they're on-hand for treating burns.

CARE

Like other succulents, aloe must be grown in a well-drained compost, such as cactus compost, under filtered (rather than direct) sunlight. In the winter, when the days shorten, aloe vera should be moved to a south-facing window. It can be moved deeper into the room in high summer. On warm days, open the windows to give your aloe plenty of fresh air.

DID YOU KNOW? GEL FROM ALOE VERA LEAVES HAS ANTISEPTIC PROPERTIES AND IS BELIEVED TO HAVE BEEN USED FOR CENTURIES AS A BALM TO TREAT BURNS.

These plants grow in arid environments in the wild so you can easily kill them with kindness. From spring to autumn, water the compost only when it gets dry, and don't be tempted to mist the plants with water. In winter, water even less frequently. You'll know if your plant is happy as it will begin to produce baby offsets that can be simply removed and potted on. If you've met all of its needs, it may produce tall clusters of orange, tubular flowers in spring, though this is rare in cooler climates.

PROBLEM SOLVING Most problems with aloes are caused by incorrect watering. Too much water and the leaves will rot at the base, too little and you'll notice brown spots on the foliage. Leaves can suffer from shock and look generally unhealthy if watered with very cold water. Try tepid water instead.

DISPLAY

Succulents require very little water for much of the year and many remain compact in pots, so you can have fun planting them in quirky containers such as large glass jars. *Echeveria* species (see p.83) have a flat habit, making them a perfect contrast to the sword-like foliage of the aloe.

ALSO TRY

There are more than 500 species in the genus *Aloe*, some of which are towering giants – *A. marlothii* can reach a height of 1.5m (5ft). These smaller, fun aloes are worth a try if you find them for sale:

- **Partridge breast aloe** (*Aloe variegata*), height 30cm (12in). In summer, this plant's rosettes of variegated foliage are sometimes offset by salmon-red flowers.
- **Spider aloe** (*Aloe humilis*), height 30cm (12in). This dwarf plant has silvery-green foliage covered with tooth-like ridges.

Partridge breast aloe is named after the pattern on the bird.

VARIEGATED PINEAPPLE

ANANAS COMOSUS VAR. *VARIEGATUS*

This eye-catching plant is grown for its spectacular pink and purple flowers and variegated, strappy leaves, although you should watch out for their sharp edges. The variegated pineapple is perfect for sunny rooms – the hotter the room, the more likely it is your plant will produce a flower spike.

HEIGHT 60cm (24in)
SPREAD 90cm (36in)
FLOWERS Pink and purple on one tall spike
FOLIAGE Variegated with spiny edge
LIGHT Sun
TEMPERATURE 16–29°C (61–84°F)
CARE Fairly easy
PLACE OF ORIGIN South America
WARNING! Leaves are serrated; gloves required when handling

The variegated pineapple thrives in a very hot room such as a conservatory.

Glorious pink and purple flowers are held on a single stalk.

DID YOU KNOW? PINEAPPLES WERE HIGHLY PRIZED IN THE 18TH AND 19TH CENTURIES, WHEN THEIR FRUITS WERE DISPLAYED ON DINING TABLES AS SYMBOLS OF STATUS AND POWER.

DISPLAY

Create an attractive grouping by placing your variegated pineapple with plants that look different but enjoy similar conditions. The succulent money plant (see p.71), for example, has the same light and water needs and its fat, round leaves will offset the long, straight foliage of your pineapple. Both of these plants are hard to kill, as is the sun-loving bunny ears cactus (see p.115).

CARE

If you have a hot, bright, sunny room, this heat-loving plant is a terrific choice. As with all bromeliads, water into the crown of the plant as well as the compost. The root system is quite small, so avoid overpotting (putting in a pot that's too large), which would cause the roots to be enveloped by a mass of moist compost. Choose a weighty pot as plants can get top-heavy.

Grow your bromeliad in an equal mix of multipurpose and orchid composts. Be sure to water plants well in spring and summer and mist them twice a week, but significantly reduce watering in winter. Applying a balanced liquid fertilizer in spring and throughout summer will improve the likelihood of a flower being produced.

Fruits will form if the pineapple gets hot enough but, unfortunately, these are bitter and inedible. Plants slowly die after the fruits have formed, but there will be offshoots at the base of the plant to pot on.

PROBLEM SOLVING The foliage can suffer from brown tips if the air is too dry or the plant has been scorched by direct sunlight. Try not to overwater as this can cause rotting of the crown. Pest problems are rare, but if you're unlucky, your pineapple plant might suffer from mealybugs (see p.46).

ALSO TRY

If you like the form and shape of variegated pineapple, these other bromeliads may also appeal:
- **Edible pineapple** (*Ananas comosus*), height 60cm (24in). This pineapple has the same care needs as the variegated pineapple and is also a great novelty plant.
- **Flaming sword** (*Vriesea splendens*), height 60cm (24in). The dark green foliage of this plant is offset by stripes. Its flower spike is stunning.

TAIL FLOWER *ANTHURIUM ANDRAEANUM*

The long-lasting flowers of this tropical beauty look spectacular against its large, arrow-shaped, dark green leaves. Its spike of small flowers is surrounded by a red spathe, making this plant a stylish choice that's guaranteed to add drama and interest to your home.

HEIGHT 45cm (18in)
SPREAD 30cm (12in)
FLOWERS Small spike, surrounded by waxy, red spathe
FOLIAGE Large, arrow-shaped
LIGHT Filtered sun
TEMPERATURE 16–24°C (61–75°F)
CARE Challenging
PLACE OF ORIGIN Columbia and Ecuador

The spikes of flowers and their red spathes can appear any time of year.

CARE

The tail flower, also known as flamingo flower, is a challenging plant to grow, but, if properly cared for, will add style to any interior. For success, place the plant in a room with both filtered sunlight and moderate humidity. The warmer and brighter the room, the greater are the chances you'll be rewarded with long-lasting blooms.

Make sure you don't ignore this slow-grower: it needs a moist compost throughout the year, which could mean you'll be watering every other day. Mist the plant every couple of days and if you're going away, put the container on a tray of damp pebbles.

Moistened moss, placed on the compost surface, will raise humidity.

To encourage flowers, give the tail flower a high-potash feed in spring with and an equal mix of loam-based and multipurpose composts.

Plants can be easily divided in spring at the same time as repotting. Keep the crown of the plant just above soil level as it will rot if planted too deep.

PROBLEM SOLVING High humidity is essential for this plant to thrive. If the air in the room is too dry, leaves will turn brown and curl and flowers won't develop. To prevent this, avoid placing it in rooms with open fires, mist plants regularly with tepid water, and cover the compost surface with moist moss.

DISPLAY

Anthuriums, rose grapes (see p.108), and birds of paradise (see p.135) work well both as cut flowers and houseplants. All enjoy similar growing conditions. Display them together as houseplants to create a dynamic tropical beach-look in your home.

ALSO TRY

The tail flower's waxy, red spathe and flower spike create a bold statement. For other plants with spathes, why not try:

- **Calla lily** (*Zantesdeschia aethiopica*), height 80cm (32in). This semi-evergreen can be grown outside if protected from frost, but it will survive indoors, as long as it's given a daily soaking during the growing season.
- **Golden calla lily** (*Zantedeschia elliottiana*), height 60cm (24in). The bright yellow, trumpet-shaped flowers of this eye-catching plant are offset by dark green foliage with white markings.

The calla lily has a white spathe and a spike of yellow flowers in summer.

CAST IRON PLANT

ASPIDISTRA ELATIOR

HEIGHT 60cm (24in)
SPREAD 60cm (24in)
FLOWERS Rare
FOLIAGE Dark green
LIGHT Light shade/shade
TEMPERATURE 7–24°C (45–75°F)
CARE Easy
PLACE OF ORIGIN China and Vietnam

The cast iron plant is slow-growing and well known for its ability to cope with neglect and extremely low light levels, which makes it ideal for north-facing rooms. It was hugely popular in Victorian times because it thrived in dark and draughty houses. This is a great plant for beginners.

CARE

This robust, undemanding plant will tolerate almost anything other than complete darkness or direct sunlight. Unlike many houseplants, it can also cope with dark, dingy corridors as well as a draughty spot by the front door.

Grow your cast iron plant in a large container in multipurpose compost. Water it regularly from spring to autumn, but reduce watering during winter months. This plant tolerates dry air, but will thrive more, and look much better, if you clean its leaves with a damp cloth to remove dust and dirt.

The cast iron plant dislikes being repotted too often, hence the need for placing it in a larger container – every three years is more than acceptable for this task. As it grows very slowly, it takes a while to fill a pot. If repotting, choose springtime, when the plant can be more easily be divided.

It's rare for an aspidistra to flower but its maroon blooms, although insignificant, are intriguing.

PROBLEM SOLVING Common causes of failure with this virtually bullet-proof plant are overwatering and too much light. Red spider mites and mealybugs can also cause issues (see pp.46–47).

DISPLAY

There are few other plants that will combine with the cast iron plant due to the need for tolerance of low light: Chinese evergreen (see p.52) and the Swiss cheese plant (see p.109) are two examples. The former also has shiny, dark leaves but a contrasting leaf shape and growing habit. The latter will lend brightness to the grouping, thanks to its variegated foliage. Add some sparkle to dark rooms or corners by choosing white or silver pots for these plants.

Cast iron plants will always look fresh if you clean their leaves with a damp cloth.

The variegated bar-room plant has leaves with irregular variegation.

ALSO TRY

Slightly more dramatic forms of the cast iron plant are also available; these have the same habits but aren't quite as tolerant of harsh growing conditions. Look out for:
- **Spotty aspidistra** (*Aspidistra sichuanensis* 'Spotty'), height 30cm (12in). Although seldom seen, this plant is well worth tracking down. Gorgeous golden spots cover its dark green, glossy leaves.
- **Variegated bar-room plant** (*Aspidistra elatior* 'Variegata'), height 60cm (24in). This plant has attractive, irregularly striped, upright foliage and flourishes in west- and east-facing rooms.

BIRD'S NEST FERN

ASPLENIUM NIDUS

Unlike a typical fern, this striking plant has undivided leathery fronds that have a very shiny surface. Its compact vase shape gives it great appeal when viewed from any side or from above. The polished, evergreen leaves catch the light, which adds to this stylish plant's considerable charm.

HEIGHT 60cm (24in)
SPREAD 40cm (16in)
FLOWERS None
FOLIAGE Wide, undivided fronds
LIGHT Light shade
TEMPERATURE 13–24°C (55–75°F)
CARE Fairly easy
PLACE OF ORIGIN Hawaii

CARE

A humid room, such as a bathroom with a frosted window or skylight, is ideal for this fern, whose native home is tropical woodland. These origins also explain why the bird's nest fern is best planted in multipurpose compost that should never be allowed to dry out.

It might be tempting to leave your plant sitting in a dish of water, but it's important to avoid this as it could cause it to become disease-prone (see below). However, a light misting twice a week is sensible if the air in the room is too dry. Apply half-strength balanced feed monthly in spring and autumn. Cleaning the leaves of the fern with a damp cloth not only keeps the plant looking tidy, but enables it to receive more light.

Repot every other year in spring and if you fancy creating more plants for free, then have a go at collecting the tiny spores from the back of the leaves. Place them on a bed of compost and cover with a sheet of glass. Keep the pot in a warm, shady spot.

PROBLEM SOLVING Given the right conditions, this is a problem-free plant, but if watered excessively it is prone to root rot. Scale insects (see p.47) can also be an issue, but don't mistake the spores of the fern for the pests as they look similar and are both found on the underside of the fronds.

Bird's nest fern enjoys high humidity and is at home in a steamy bathroom.

This plant's spores are to be found on the underside of its fronds.

DISPLAY

The bird's nest fern, with its arching fronds, is well-suited to a hanging basket. This is a great way of displaying the plant in a bathroom, where surface space can be limited. For a tropical display with impact, pair this fern with other types of fern, such as the maidenhair fern (see p.50) and Boston fern (see p.112), both of which look equally at home in a hanging basket.

ALSO TRY

The glossy, bright green foliage of the handsome bird's nest fern adds style to both traditional and contemporary interiors. This is an extremely tidy plant, with an unfussy appearance. Other similarly stylish plants include:

- **Boat lily** (*Tradescantia spathacea*), height 45cm (18in). The boat lily is perfect for a shady room that has high humidity. It has neat, green and purple sword-shaped leaves that point upwards.
- **Pleated bird's nest fern** (*Asplenium nidus* 'Crispy Wave'), height 60cm (24in). This plant shares all the attributes and care needs of the bird's nest fern but has attractive, ruffled leaves.

PONYTAIL PALM

BEAUCARNEA RECURVATA

As soon as you see this plant you'll understand why it's called the ponytail plant: it forms a fountain of long, thin, strap-like leaves that make it a unique upright specimen and a frequent talking point. This easygoing, elegant plant is simple to place as it enjoys both filtered sunlight and low humidity.

HEIGHT 2m (6ft 6in)
SPREAD 1m (3ft 3in)
FLOWERS Unlikely to bloom indoors
FOLIAGE Thin, strap-like
LIGHT Sun/filtered sun
TEMPERATURE 10–26°C (50–79°F)
CARE Easy
PLACE OF ORIGIN Mexico
WARNING! Leaves are slightly serrated; gloves require when handling

The ponytail palm is slow-growing but extremely long-lived.

CARE

When fully mature, this unique plant is a wonderful addition to a well-lit entrance hall, where it looks best planted in a large, floor-standing pot. One of the plant's most distinctive features is its trunk, which has a swollen base that enables it to store water. When placing it in your home, bear in mind it's fun to showcase this feature.

Although labelled as a palm, the ponytail plant is in fact more closely related to the yucca (see p.140). Like the yucca, it's tough, thrives on neglect, and prefers being planted in a cactus compost. The bigger the pot you plant it in, the larger your plant will grow.

Water sparingly and feed in spring with a half-strength, balanced fertilizer. Avoid draughts, especially in winter. The long leaves will benefit from being cleaned with a damp cloth, but watch out for the slightly serrated leaf edges.

Flowers are rare when grown as a houseplant, as are the offshoots that can appear at the base of the adult plant.

PROBLEM SOLVING A common problem is stem rot – caused by excess water – which can kill a plant. You can try to remedy this by dramatically reducing watering, but it's best to avoid the issue by not overwatering in the first place. As with many plants placed in warm, bright rooms, your plant may attract red spider mites (see p.47 for solutions). Reduce the chances of encountering these pests by misting the plant.

DISPLAY

The foliage of the ponytail palm tumbles towards the floor. For dramatic effect, partner it with plants that have foliage that does the opposite but share the same care needs. The variegated snake plant (see p.128), for example, displays upright, sword-shaped foliage and reaches 75cm (30in) in height. The bird's nest sansevieria (see p.127), which reaches just 20cm (8in) in height, is the perfect partner for smaller, tabletop-sized ponytail palms.

ALSO TRY

The ponytail palm's ability to store water and cope with irregular watering makes it ideal for people who are often away from home. If this appeals, then also consider:

• **Chandelier tree** (*Euphorbia triangularis*), height 1m (3ft 3in). This attractive cactus look-alike has a ribbed and thorny stem.

• **Peruvian cactus** (*Cereus*), height up to 90cm (36in). This plant has the familiar and classic cactus shape, with spiny stems and pink or white flowers.

The chandelier tree is an attractive, undemanding succulent.

BEGONIA *BEGONIA*

Begonia leaves are decorative and textured, with some variegated, spotted, or attractively veined. There are also numerous foliage shapes to choose from and the flowers add a rainbow of colours. Display wherever you can best admire the foliage patterns of these lovely trailing or upright plants.

HEIGHT Up to 90cm (36in)
SPREAD Up to 45cm (18in)
FLOWERS Numerous, depending on variety
FOLIAGE Variegated, patterned
LIGHT Filtered sun/light shade
TEMPERATURE 15–22°C (59–72°F)
CARE Fairly easy
PLACE OF ORIGIN Central and South America and Asia

The Rex begonias offer the most dramatic foliage in the entire group.

CARE

To appreciate their intricate foliage, place these plants on a raised shelf. An east-facing room is perfect, but in the heat of summer, you'd be wise to move them further back into the room, away from any windows, to avoid foliage scorching. Moderate humidity is ideal, which makes a kitchen the perfect spot.

Grow begonias in multipurpose compost and sit your pot on a tray filled with expanded clay granules; this must be kept moist and will increase humidity around the plant. Do not mist the leaves, as this can cause mildew.

Keep the plant compact by pinching out any leaves that are spoiling the silhouette. Don't be tempted to clean begonia leaves as they are easily damaged. Give a high-potash feed in the summer to encourage flowers. Once flowers have faded, deadhead (remove) them to keep the plant tidy and encourage further blooms.

Propagate popular houseplant begonias such as the spectacular *Begonia rex* (see *left*) by taking leaf cuttings in May (see *p.38*). Adventurous gardeners might want to try to grow plants from tubers in March or April.

PROBLEM SOLVING Brown leaf tips, a common problem with begonias, are caused by low humidity, scorch, or plants being placed too close to a heat source. Rotting leaves are caused by overwatering or too much shade. Damaged leaves should be removed, or the brown tips snipped off with scissors.

DISPLAY

Begonias look spectacular when displayed as a group. Plant them in individual pots placed on a large, decorative tray filled with damp, expanded clay granules. Another option is to place African violets (see *p.126*) and Cape primroses (see *p.136*) on the same tray with them, as these flowering plants have the same needs.

ALSO TRY

Few plant groups offer such variety as the begonia when it comes to colourful flowers and foliage, but these alternatives come close:

- **Busy Lizzie** (*Impatiens* 'New Guinea' hybrids), height 30cm (12in). These popular plants enjoy the same conditions as begonias and offer a profusion of pink, lilac, red, or white flowers over deep green or variegated foliage.
- **Earth star** (*Cryptanthus*), height 15cm (6in). This diverse genus of bromeliads is grown for its stunning foliage: some plants have pink, yellow, and green leaves, others offer snakeskin patterns.

Earth star is available in a range of magnificent and vibrant colours.

FRIENDSHIP PLANT

BILLBERGIA NUTANS

This popular bromeliad thrives on neglect. It offers a rosette of grey-green foliage that will arch elegantly over the side of a container. It's grown primarily for its attractive salmon-pink stems and bracts and its multicoloured flowers. The friendship plant thrives in a room with high humidity.

HEIGHT 60cm (24in)
SPREAD 60cm (24in)
FLOWERS Multicoloured
FOLIAGE Strap-like, grey-green
LIGHT Filtered sun/light shade
TEMPERATURE 16–27°C (61–81°F)
CARE Easy
PLACE OF ORIGIN Brazil

CARE

For success, the friendship plant requires a room with light shade or filtered sunlight as this will help to encourage the tubular green, yellow, pink, and purple blooms.

Plant in an equal mix of orchid and multipurpose compost. Water with rain or distilled water and allow the compost to dry out between watering in the spring and summer; reduce watering in autumn and winter. The friendship plant will thrive on a regular misting. Apply a half-strength, balanced fertilizer to the plant every month during spring and summer.

This epiphyte will take a few years to flower, but it's worth the wait. Blooms can appear at any time of year, but most commonly in summer; they last well over a month. After flowering, the shoot the flower came from will die, but there will be plenty of new shoots to take over. The plant reliably produces offsets, so it quickly fills a generous pot. When repotting in spring, remove the offsets and pot on.

PROBLEM SOLVING This easy-to-grow plant rarely has problems with pests and diseases. If growing in a very hot conservatory or greenhouse it might be affected by red spider mites (see p.47).

A room with very low humidity is likely to cause the tips of the leaves to go brown. Remedy by misting regularly.

DISPLAY

Display this elegantly arching plant in a position where it can be viewed from all sides. The friendship plant looks at home with other bromeliads such as the scarlet star (see p.94) and the blushing bromeliad (see p.111). All need high humidity, filtered sun, and well-draining compost, so place them together in one large planter. None of them is deep rooted, so they could share a shallow, decorative container.

The friendship plant has striking, salmon-coloured stems and bracts.

ALSO TRY

Plants with arching habits are perfect for displaying on a plinth or at the centre of a small table. Another plant that requires little in the way of maintenance and has this attractive growing habit is:

• **Fishbone cactus** (*Epiphyllum anguliger*), height/trail 60cm (24in). Unlike the friendship plant, this easy-care cactus requires low humidity and bright sunlight. The foliage starts to arch attractively as the plant matures. It has unusual flat, green stems.

PAPER FLOWER

BOUGAINVILLEA GLABRA

If given enough support, this showy climber will quickly decorate your room with magnificent, floor-to-ceiling colour, thanks to its papery bracts. This is the ideal plant for a hot conservatory – it will flourish in the sunlight while also providing you with some shade.

HEIGHT Up to 6m (20ft)
SPREAD Up to 1.5m (5ft)
FLOWERS Pink, red, or white
FOLIAGE Small, green
LIGHT Sun
TEMPERATURE 10–26°C (50–79°F)
CARE Fairly easy
PLACE OF ORIGIN South America
WARNING! All parts are toxic; gloves required when handling

CARE

The paper flower enjoys heat, bright light, and low humidity, which makes it a winning plant for a south-facing room or conservatory. If you give it a trellis to climb up, it will soon cover a wall, but you can keep it smaller by pruning to size and training it over a hoop.

If you want your plant to climb, put it in a sizeable, sturdy container with a loam-based compost. By growing it in a large pot you can avoid repotting it every spring – simply add a fresh layer of compost to the top of the pot.

Add slow-release, balanced, granular feed in spring. Water well from spring to autumn; reduce watering in winter.

Prune in late winter as the flowers are produced on the new season's growth. If you prune in early summer you're in danger of missing out on the wonderful paper bracts that surround the tiny central blooms. You can propagate this spectacular tropical climber by taking cuttings in summer.

PROBLEM SOLVING When paper flower plants are grown in a very hot conservatory their leaves may be scorched by the direct sunlight. Avoid this, if possible, by adding shading to the structure in high summer. If the temperature of the conservatory drops below 10°C (50°F) in winter, the leaves may drop. The plant could be attacked by mealybugs, but if you spot the white cottony fluff early enough, wipe them off with a damp cloth (see p.46).

DISPLAY

Living walls and climbers are becoming more popular as they give big impact and take up very little floor space. Allow your paper flower to share a wall or trellis with the magnificently scented jasmine (see p.103) or the Cape leadwort (see p.124) and, if they flower all at once – which is very possible – you'll have a truly enviable display.

Tiny white flowers offset this plant's spectacular pink bracts.

ALSO TRY

Consider partnering your paper flower with some equally stunning climbers, such as passion flowers, if you're planning to cover your walls. These tropical-looking plants will grow up a trellis but can easily be kept more compact by pruning. Good options include:

- **Passion flower** (*Passiflora caerulea*), height 4m (13ft). This intricate and eye-catching plant enjoys the same hot conditions as the paper flower.
- **Red passion flower** (*Passiflora racemosa*), height 4m (13ft). This is a more tender and challenging plant than the paper flower but has wonderful red blooms in summer, followed by fruits.

If space is an issue, train this plant over a hoop and keep it clipped to size.

ANGEL'S TRUMPET

BRUGMANSIA × *CANDIDA*

When in flower, angel's trumpet is an absolute showstopper. The plant's pendent, trumpet-shaped, evening-scented blooms can grow as long as 20cm (8in). Due to its size, it makes the perfect plant for a spacious conservatory or a sunny room adjoining a patio.

HEIGHT Up to 2m (6ft 6in)
SPREAD Up to 1m (3ft 3in)
FLOWERS Large, scented, trumpet-like
FOLIAGE Ovate, green
LIGHT Sun/filtered sun
TEMPERATURE 16–25°C (61–77°F)
CARE Fairly easy
PLACE OF ORIGIN South America
WARNING! All parts are toxic; gloves required when handling

CARE

In summer, this is a terrific plant for a south-facing room with an adjacent patio – its scented flowers can be overwhelming in the evening and are best enjoyed outside. But in winter, a sunny room is the perfect spot.

Grow in a large container, otherwise the plant – which is often trained into a tree or a standard – may become top-heavy. A heavy, loam-based compost will ensure it stays upright.

Angel's trumpet needs plenty of feed and water throughout the spring and summer, so apply a balanced liquid fertilizer in spring but then swap to a high-potash feed in summer to bring on the flowers. Water well until winter, when watering should be reduced.

The flowers of the angel's trumpet can be an impressive 20cm (8in) long.

DID YOU KNOW? THE FLOWERS OF THIS PLANT ARE SCENTED AT NIGHT AND, IN THEIR NATIVE SETTING, ATTRACT HUMMING BIRDS.

Once the flowers have faded, prune your plant to keep it in shape. Look out for suckers at the base of the plant and cut them off as soon as they appear. Be sure to wear gloves when handling or pruning this plant.

PROBLEM SOLVING Angel's trumpet will soon droop if it's planted in too small a pot or if it's underwatered. This is easily solved by repotting the plant in spring and watering more generously in the summer.

Red spider mites (see p.47) are often attracted to this plant; severe infestations will cause speckled foliage. Misting may help reduce the problem.

DISPLAY

Angel's trumpet will look at home in a large container, but consider investing in a plant trolley if you aim to move it in and out of the house.

Pair your plant with other sun-lovers such as the spineless yucca (see p.140) – this will be similar in height yet offer a contrast in shape. If growing in a large pot, plant a spider plant (see p.65) under the angel's trumpet as it will also enjoy a summer outside.

ALSO TRY

If you've fallen for plants, such as angel's trumpet, with striking blooms, try these challenging beauties:

- **Chinese lantern** (*Abutilon* × *hybridum*), height 90cm (36in). This plant likes filtered sunlight and has bell-shaped flowers in a variety of colours throughout summer.
- **King's crown** (*Justicia carnea*), height 1.2m (4ft). Although rarely seen, this plant is worth searching for. It has oblong leaves and spikes of pink, tubular flowers.

Flowering maple likes a room with filtered sun and low humidity.

HEARTS ON A STRING

CEROPEGIA LINEARIS SUBSP. *WOODII*

This delicate yet easygoing trailer cascades from shelves and hanging baskets, creating a waterfall of heart-shaped, silvery leaves. It's an ideal plant if space is tight and a hanging garden is your only real option. Display it in a bright, sunny room with moderate humidity such as a kitchen.

HEIGHT Up to 5cm (2in), trailing to 90cm (36in)
SPREAD 20cm (8in)
FLOWERS Small, pink, tubular
FOLIAGE Tiny, patterned, heart-shaped
LIGHT Sun/filtered sun
TEMPERATURE 18–24°C (64–75°F)
CARE Easy
PLACE OF ORIGIN South Africa

CARE

With its long, elegant, trailing stems, this plant needs room to tumble, which makes it perfect for shelves and hanging baskets. It thrives in moderate humidity and a sunny room but keep it away from direct sunlight, which can cause the colourful leaves to fade.

The foliage of hearts on a string is succulent, so grow in cactus compost and keep it fairly dry. Moisten lightly in spring and summer, but reduce the watering in winter. It doesn't require much in the way of feeding, so apply a balanced fertilizer once a month during the spring and summer.

This plant is grown primarily for its delicate trailing foliage, but it will, on occasion, produce small, pink flowers in summer. After the flowers have faded, you'll notice small nodules on the stems. New plants can be formed by cutting off a section of stem – when the nodules make contact with the compost, roots will eventually form and create a new plant.

PROBLEM SOLVING If you get the watering and light levels right, you shouldn't have any problems with this plant. Underwatering will cause brown spots to appear on the leaves or the leaves to drop. If they start to look droopy and lose colour, this is probably due to overwatering. The plant is relatively pest- and disease-free.

DISPLAY

With its delicate, somewhat romantic look, this plant is perfect for a pastel interior and is particularly striking in rooms that are decorated in pinks, blues, silver, or white. Hearts on a string partners extremely well with other hanging or trailing plants – such as the creeping fig (see *p.90*) or inch plant (see *p.139*) – in baskets that are hooked to a supporting beam.

Hearts on strings have beautifully patterned leaves with a silvery shimmer.

String of pearls tumbles from a basket to meet neighbouring plants.

ALSO TRY

You can create a foliage waterfall by grouping trailing plants that are similar to hearts on a string, such as:

- **Mistletoe cactus** (*Rhipsalis baccifera*), height 30cm (12in), trails to 1.8m (6ft). Despite its common name, this plant likes high humidity and indirect sunlight. Handle with care as the stems break easily.
- **String of pearls** (*Curio rowleyanus*), length 90cm (36in). With its tiny, fleshy, bead-like leaves, this plant shares the same care needs as hearts on a string.

PARLOUR PALM

CHAMAEDOREA ELEGANS

This elegant, adaptable houseplant can be grown in a bright spot or in shade, hence its undying popularity. It's easy to place in any home and excellent for purifying air. This is the perfect plant to buy as a gift for someone whose house you've never visited before – you just can't go wrong.

HEIGHT 1.2m (4ft)
SPREAD 60cm (24in)
FLOWERS Unlikely to bloom indoors
FOLIAGE Slender, palm-like
LIGHT Filtered sun/light shade
TEMPERATURE 15–27°C (59–81°F)
CARE Easy
PLACE OF ORIGIN Mexico

CARE

The parlour palm is one of the most popular houseplants. Immensely robust and tolerant, it will thrive in any room, cool north or hot south, and copes with low to moderate humidity. Its upright, feathered foliage offers a classic look that suits almost every style of home.

Choose a generous container for your parlour palm and plant it in a multipurpose compost. Allow the compost to dry out slightly between watering in summer; water sparingly during the winter months. Apply a balanced liquid feed between spring and autumn, but stop feeding in winter.

It's quite normal for a whole frond to sometimes turn brown – if this happens, simply cut it off at the base. Parlour palms are slow-growing plants but, in time, new leaves will form. It will only need repotting every two years, thanks to this slow rate of growth.

It's rare to see this plant bloom, but when it does, the panicles of small, yellow flowers aren't really worth the wait as they tend to be quite scruffy.

PROBLEM SOLVING Although mostly problem-free, this palm can attract thrips, scale insects, and red spider mites (see p.47). The tips of leaves will turn brown if placed in a draughty spot. Remove these with scissors and place your plant somewhere warmer.

DISPLAY

This eye-catching architectural plant looks best planted alone in a container. Choose a pot that will complement the style of your interior, and the elegant qualities the plant has to offer. Group it, in matching pots, with other impressive palms, such as sago palm (see p.73) and kentia palm (see p.100). Create contrast and drama in your display by choosing three palms of different heights.

The parlour palm is the top pick for a contemporary home with shady rooms.

ALSO TRY

Palms offer height, reliability, and splendour. If this is the look you're after, then try some of the lesser-known types – all will look distinctive in the corner of a room:

- **Areca palm** (*Dypsis lutescens*), height 2m (6ft 6in). This neat palm is perfect for a tight corner. Its leaves are slightly glossier than those of the parlour palm.
- **Fishtail palm** (*Caryota mitis*), height 2.5m (8ft). The serrated foliage of this palm resembles a fishtail. It requires slightly higher humidity than the parlour palm.

The areca palm needs plenty of space as its foliage arches dramatically.

SPIDER PLANT

CHLOROPHYTUM COMOSUM

The spider plant is one of the best-known and most popular houseplants, and with good reason: not only is it easy to grow but it will tolerate a fairly cool or a warm room and looks spectacular tumbling gracefully over the edge of containers and hanging baskets.

HEIGHT 30cm (12in), trailing to 40cm (16in)
SPREAD 40cm (16in)
FLOWERS Insignificant
FOLIAGE Flowing, variegated
LIGHT Filtered sun/light shade
TEMPERATURE 7–24°C (45–75°F)
CARE Easy
PLACE OF ORIGIN South Africa

A mature spider plant with a healthy crop of plantlets.

CARE

Spider plants are the perfect starter plant for beginners. They flourish in indirect sunlight or light shade and will grow best in a room with low humidity. Plant them in multipurpose compost and keep moist in spring and summer but allow them to dry out between watering during the winter months.

Apply a balanced liquid fertilizer to your spider plant twice a month throughout the spring and summer. Repotting should only be necessary every two to three years.

Placing plantlets in jars of water will encourage rooting.

Small, white flowers appear on the end of the wiry stems; these are often referred to as runners and are home to plantlets. Propagation is quick and easy: peg down a plantlet in a pot of compost or place in a jar of water, cutting it free from the parent plant once it has rooted (two to three weeks). Alternatively, simply leave the plantlets attached to the parent plant, where they will add to the display.

PROBLEM SOLVING Leaves will turn brown if plants are overwatered or cold in winter months. Limp and yellow leaves are usually caused by excessive heat and insufficient light. Mature plants should produce flowers and plantlets, but if these fail to appear, encourage them by planting in a fairly tight pot. Spider plants are only rarely, if ever, attacked by pests.

DISPLAY

The arching foliage looks spectacular in a hanging basket, pot, or at the edge of a container with other plants, and will brighten even the gloomiest room. If a hanging basket isn't possible in your house due to space or other practical reasons, put your plant on a pedestal or shelf to tumble. Its bright leaves will mix well with dark foliage plants, so place it on a pedestal next to taller plants such as a rubber plant (see p.88) or a Swiss cheese plant (see p.109), or plant it at the base of them in the same pot.

ALSO TRY

Hanging baskets allow you to fill a room with flowers and foliage when floor space is limited. Other great trailing plants include:

- **Lipstick plant** (*Aeschynanthus pulcher*) trails to 30cm (12in). This plant thrives in a room with filtered sunlight and moderate humidity. Its dark green foliage is offset by tubular, red flowers.
- **Purple spiderwort** (*Tradescantia pallida* 'Purpurea'), height 30cm (12in), trail 60cm (24in). The standout feature of this plant is stunning purple foliage with a dramatic, pointed shape.

GRAPE IVY *CISSUS RHOMBIFOLIA*

The grape ivy has been a popular houseplant for decades, thanks to its versatility. Its long stems of dark green leaves can be trained up a trellis or left to tumble over the side of a basket. A room with low humidity and filtered sunlight is where this vine will be most content.

HEIGHT 30cm (12in), trailing or climbing to 1.5m (5ft)
SPREAD 60cm (24in)
FLOWERS Unlikely to bloom indoors
FOLIAGE Green, lobed
LIGHT Filtered sun/light shade
TEMPERATURE 12–24°C (54–75°F)
CARE Easy
PLACE OF ORIGIN South America

Grape ivy is a versatile plant that can be trained to grow up or left to tumble.

CARE

If you're keen to cover a wall, trellis, or pole with foliage, this is an ideal plant as it self-climbs, clinging onto supports and surfaces by means of aerial roots. Grape ivy thrives in low humidity, so it's better suited to a sitting room than to a kitchen or bathroom.

Specimens grow to fill the size of their containers, so if you want to cover an entire wall with foliage, plant the grape ivy in a generous pot that's filled with loam-based compost. This is a plant that likes good aeration at the roots, so mix some chipped bark into the compost when repotting in spring.

Feed your plant with a balanced fertilizer once a month throughout spring and summer and keep the compost moist. Reduce watering in winter. However, no amount of feed will entice the grape ivy to bloom – flowers will rarely appear.

To encourage a dense plant, lightly prune in late summer. Simply cut off the stem tips at a leaf joint.

PROBLEM SOLVING Grape ivy is not prone to pest infestation, but mildew on leaves can become an issue if the compost is poorly drained. Remedy this by repotting, removing diseased leaves, and improving the ventilation around the plant. Spotted or curled leaves are frequently the result of compost that's too dry.

DISPLAY

Grape ivy isn't a fast-growing plant, but if pruned it will, in time, cover a trellis to produce a dense and impressive screen of living foliage, which can be used as a very effective room divider. To create this effect, simply plant your grape ivy in a trough alongside other houseplant climbers, such as devil's ivy (see p.84) and the waxflower (see p.101).

ALSO TRY

If adding interest to walls with live plants appeals to you, there's plenty of choice. Experiment with these more unusual climbers up an obelisk or trellis in your home:

• **Firecracker plant** (*Manettia luteorubra*), height/trail 1.8m (6ft). This plant has pointed, dark leaves and red, tubular flowers that are tipped with yellow during the summer months.
• **Variegated Natal ivy** (*Senecio macroglossus* 'Variegatus'), height/trail 1.5m (5ft). Although often mistaken for the common ivy, this plant has fleshier, glossier foliage.

Variegated Natal ivy's foliage combines well with plain-leaved plants.

CITRUS TREE *CITRUS*

Growing your own orange or lemon tree is an exciting prospect, and one that's achievable if you have a well-lit room or a conservatory. These plants are tricky to please, but care and attention will be amply rewarded – they flower nearly all year long and are never short on interest.

HEIGHT 1.5m (5ft)
SPREAD 1m (3ft 3in)
FLOWERS White, scented
FOLIAGE Dark green
LIGHT Sun
TEMPERATURE 13–24°C (55–75°F)
CARE Challenging
PLACE OF ORIGIN South Asia

Lemons prefer to ripen on the tree, so only harvest them when they're yellow.

CARE

Citrus plants are self-fertile, so you don't need to grow any more than one plant to enjoy the oranges or lemons from mature specimens.

Although these are quite challenging plants to grow and care for, they will thrive in a bright south-facing room that's warmed naturally by the sun, rather than one with central heating,

Choose the sunniest spot in the house to encourage a healthy crop.

which is far too drying for them. However, it's advisable to grow your tree in a specially formulated citrus compost and also to feed it with a specialist citrus fertilizer. If you can't get hold of the right compost, then mix a handful of horticultural grit into the growing medium, which will encourage good drainage.

These plants will also flourish if they're placed in a humid area and watered freely during the summer months with tepid rainwater. Regular misting helps raise humidity levels around the plants, as does sitting their containers on a tray of moist, expanded clay granules.

Pruning requirements are minimal and straightforward – in February, clip your citrus tree to the desired shape.

PROBLEM SOLVING Plants that enjoy extremely warm temperatures, such as citrus, are always more prone than other plants to pests, including mealybugs, red spider mites, and scale insects (see pp.46–.47). Regularly check the undersides of leaves, where pests lurk, and move infected plants away from all other houseplants.

DISPLAY

Large citrus plants look grand and imposing when displayed in sizeable terracotta or glazed pots. Smaller specimens are striking if grouped with other sun-loving plants such as rose grape (see p.108).

ALSO TRY

Other, more unusual, citrus are also available. Try kumquats and limes, for example; they can grow quite tall if given large pots and plenty of time:

- **Kumquat** (*Citrus japonica*), height 1m (3ft 3in). The orange fruits of the kumquat are followed by scented flowers. It thrives in sun.
- **Lime** (*Citrus* × *latifolia*), height 1m (3ft 3in). This plant has sharp-tasting green fruits that are followed by scented flowers.

BLEEDING-HEART VINE

CLERODENDRUM THOMSONIAE

When in flower, this climbing or trailing plant is a real showstopper. Its ideal spot is a glass porch with shade in summer or a room with filtered sunlight and high humidity. Given these conditions, the bleeding-heart vine will leap up a trellis and decorate your walls like no other plant.

HEIGHT 3m (10ft)
SPREAD 1.5m (5ft)
FLOWERS Red with white calyces
FOLIAGE Green, shiny
LIGHT Filtered sun
TEMPERATURE 12–24°C (54–75°F)
CARE Fairly easy
PLACE OF ORIGIN West Africa

CARE

To achieve the bleeding-heart vine's stunning blooms – made up of long-lasting, red flowers surrounded by dramatic, white calyces – it's important to provide the right growing conditions.

The key to success is plenty of filtered sunlight. Plant in multipurpose compost; keep this moist, but not wet, in spring and summer. Don't be tempted to overpot this plant as a tighter container will encourage more flowers, as will regular applications of

Flowers appear any time of year, but summer is most likely.

a balanced fertilizer. The bicoloured blooms can appear at any time of the year but are most prolific in summer. When sufficiently fed and watered, bleeding-heart vine is a vigorous grower. But don't let this put you off as it can be clipped to size in winter: prune it to a small size if you want to grow it as a tabletop plant or let it go free so that it will decorate a wall or tumble from a hanging basket. Cuttings from the bleeding-heart vine will root very easily if they are taken in the spring or early summer.

PROBLEM SOLVING Falling leaves in winter are normal and will soon be replaced with fresh foliage. Brown leaf tips are often due to low humidity. Remedy this by misting regularly. The

bleeding-heart vine is prone to red spider mites (see *p.47*), but the chances of an attack can be reduced by misting.

DISPLAY

For maximum impact and flower power, train this plant up a trellis or obelisk, or against a wall. Display it alongside other sun-loving climbers, such as paper flower (*p.61*) and jasmine (*p.103*), and together they'll cover a south-facing room with magnificent blooms. If you have a conservatory, run galvanized wires across the ceiling and the plant will climb among them.

Clerodendrons can be grown inside or outside during a mild summer.

ALSO TRY

If you're searching for other unusual or novelty flowers then try these striking houseplants:

- **Desert pea** (*Swainsona formosa*), height 80cm (32in). The dramatic red flowers of the desert pea resemble the claws of a lobster. This is a challenging plant to grow, but your commitment will be amply rewarded.
- **Bottlebrush flower** (*Callistemon citrinus*), height 1.8m (6ft). This plant has spikes of distinctive red blooms. It's easy to grow and copes well with dry air.

NATAL LILY *CLIVIA MINIATA*

HEIGHT 45cm (18in)
SPREAD 45cm (18in)
FLOWERS Trumpet-like clusters
FOLIAGE Dark green, strap-like
LIGHT Filtered sun
TEMPERATURE 10–23°C (50–73°F)
CARE Challenging
PLACE OF ORIGIN South Africa

This plant is grown for its impressive clusters of trumpet flowers in orange, yellow, or apricot, with up to 20 flowers per spike. In order for flowers to form, this bulbous plant needs a period of dormancy in winter. A native of woodlands in South Africa, it needs protection from direct sunlight.

Natal lilies may not flower in the first year, but your patience will be rewarded.

CARE

This plant needs two rooms – a warm room for summer and a cooler one to rest in throughout the winter months. The perfect place to encourage its magnificent flowers is a bright room that's protected from direct sunlight. If you leave your Natal lily in a bright, heated room all year long, it simply won't flower. However, you should avoid moving this plant when it's in bud – it's best to find the right location during the spring and summer months and then stick to it.

Grow in multipurpose compost, feed with a balanced fertilizer, and don't be tempted to overpot as a tight pot will encourage blooms. Water well during the spring and summer months; in winter, reduce watering dramatically, stop feeding, and allow the plant to rest until the spring. Plants can be repotted and divided at this time. When doing this, make sure that the neck of the bulb is above the soil level.

After the summer flowers have faded, cut the stem back close to the base of the plant, unless you want to save the seed for sowing.

PROBLEM SOLVING If a plant fails to flower, this is often because it's too young, has not had a period of winter dormancy, or has been overpotted. These plants rarely suffer from pests, but they will rot if overwatered. This also causes the leaves to turn yellow.

DISPLAY

Natal lilies look spectacular when placed on a table in a decorative pot. To prevent the flowers from leaning towards the light, make sure the pot can easily be turned. For a dramatic display, partner your plant with a collection of begonias (see p.59). Their dramatic foliage is a wonderful contrast to the long, strappy foliage, and they enjoy similar conditions.

Display your plant at a height that will let you peer into its gorgeous blooms.

ALSO TRY

There are several other houseplants that thrive in a bright, sunny room with filtered light. The following options aren't quite as dramatic as the Natal lily, but they'll certainly flourish and look great alongside it:

- **Cigar plant** (*Cuphea ignea*), height 30cm (12in). Red, cigar-shaped flowers make this a novelty plant. It looks best next to plants with more stature.
- **Goldfish plant** (*Columnea gloriosa*), height/trail 60cm (24in). This plant has orange, tubular flowers with waxy, trailing foliage.

CROTON *CODIAEUM VARIEGATUM* VAR. *PICTUM*

Crotons have wonderfully variegated foliage in pinks, greens, yellows, and oranges. Unsurprisingly, the vibrant and colourful plant is also known as Joseph's coat. This is a spectacular houseplant that, given the right care and conditions, will enliven any room in your home.

HEIGHT 1.2m (4ft)
SPREAD 75cm (30in)
FLOWERS Insignificant
FOLIAGE Colourful, variegated
LIGHT Filtered sun
TEMPERATURE 15–25°C (59–77°F)
CARE Fairly easy
PLACE OF ORIGIN Pacific Islands and Malaysia
WARNING! All parts toxic; gloves required

Grow your croton in a bright spot to strengthen the colour of its leaves.

Display against a neutral backdrop for the foliage to really stand out.

CARE

When mature, crotons create a woody, shrub-like form and will bring vibrancy and interest into your home. However, it's important that they're kept at a consistent temperature over 15°C (59°F); sudden fluctuations in heat levels will cause their leaves to drop. They also need high humidity to really thrive: a bathroom is the ideal location.

Plant in a generous pot in loam-based compost. To help increase humidity around the plant, place the container on a tray of expanded clay granules, which should be keep moist at all times. Misting the plant will also increase humidity levels. Make sure the compost itself is moist and feed with a balanced liquid fertilizer from spring to autumn, but reduce watering and stop feeding the plant during the winter.

Repot your croton every other year, and prune in February to keep the plant neat, but wear gloves for this task as the sap can irritate the skin. Keep the colourful leaves looking fresh by cleaning them with a damp cloth.

PROBLEM SOLVING Leaf drop is the single most common problem facing crotons. This is often due to the compost drying out, the temperature dipping, insufficient light, the air being too dry, or plants being positioned close to either a heat source or a draught. Crotons are also susceptible to red spider mites (see p.47).

DISPLAY

This plant's foliage is so colourful and spectacular that it looks best set against a neutral background. You can partner your croton with a plain green bathroom plant such as the staghorn fern (see p.123) or the bird's nest fern (see p.57).

ALSO TRY

Few foliage plants enjoy such high humidity as the croton, but some thrive in these conditions and also give a dramatic display, including:
- **Zebra plant** (*Aphelandra squarrosa*), height 60cm (24in). This plant has dark green leaves with cream stripes and unusual yellow and orange blooms throughout the autumn.

The zebra plant produces stunning bright yellow flowers in autumn.

MONEY PLANT
CRASSULA OVATA

The money plant takes its name from the belief, in some cultures, that it brings prosperity. If this doesn't encourage you to grow it, then the fact that it's indestructible might. This plant is ideal for a sunny, south-facing room with low humidity and for those who may forget to water it.

HEIGHT 90cm (36in)
SPREAD 90cm (36in)
FLOWERS Unlikely to bloom indoors
FOLIAGE Green, succulent
LIGHT Sun/filtered sun
TEMPERATURE 15–25°C (59–77°F)
CARE Easy
PLACE OF ORIGIN South Africa

CARE

Not many houseplants can cope with the heat and brightness of a sunny windowsill, but the money plant will thrive in such a spot. When mature, it resembles a bonsai tree, thanks to its thick stems, which makes it hugely appealing. Avoid placing it in a humid room such as a kitchen as this can lead to poor plant health.

Grow your money plant in a cactus compost or a loam-based compost mixed with sharp sand at a ratio of 3:1.

The money plant is ideal for anyone in search of an undemanding plant.

> **DID YOU KNOW?** THERE ARE MORE THAN 200 DIVERSE SPECIES OF *CRASSULA*, RANGING FROM MINUTE, MOSS-LIKE PLANTS TO SMALL TREES.

The money plant is a succulent and overwatering can prove to be fatal, so be sure to allow it to dry out between waterings, and also reduce watering dramatically during the winter months. This is an undemanding plant that will need feeding just a couple of times throughout spring and summer with a half-strength, balanced fertilizer.

It's unlikely your money plant will flower, so grow this succulent for its shape and red-tipped foliage. Turn the pot occasionally to encourage a neat and uniform framework. The tree-like form can make the plant top-heavy, so choose a substantial, heavy, glazed container to hold it steady.

PROBLEM SOLVING You'll soon notice if this plant isn't happy by inspecting its leaves. Brown spots are likely to be caused by underwatering during the summer, whereas wilted and yellowing leaves are usually a sign of overwatering. You can help prevent leaves dropping by watering with tepid tap water.

DISPLAY

Consider planting a window box with other sun-loving succulents and placing it on a bright, indoor sill. The money plant will combine well with an echeveria or two (see p.83) and a spiky zebra cactus (see p.96). Alternatively, plant them in individual pots that can easily be moved about the house as the mood takes you.

ALSO TRY

Many succulents cope with limited watering, which makes them good partners for crassulas and great for anyone who's often away from home. Why not try:

• **Black aeonium** (*Aeonium* 'Zwartkop'), height 60cm (24in). This plant has rosettes of dark purple foliage held on thick stems.

• **Century plant** (*Agave americana*), height 90cm (36in). With its long, spiky, striped leaves, this an impressive plant. Be careful as the foliage has very sharp spikes.

Aeoniums will also thrive outside but require frost protection.

NEVER-NEVER PLANT

CTENANTHE BURLE-MARXII

This plant will happily grow under far larger houseplant specimens. It thrives in a shady room with high humidity but must be kept out of direct sunlight, which fades its fancy, fishbone-patterned foliage. The decorative leaves will inject sparkle and interest into even the gloomiest areas.

HEIGHT 60cm (24in)
SPREAD 45cm (18in)
FLOWERS Insignificant
FOLIAGE Fish-bone patterned
LIGHT Filtered sun/light shade
TEMPERATURE 10–25°C (50–76°F)
CARE Easy
PLACE OF ORIGIN Brazil

CARE

To keep the impressive, silvery-green, oblong foliage of the never-never plant in prime condition, mist it regularly. Make sure that the compost remains moist between the spring and autumn months. Reduce watering in winter, increase it again in springtime, and start feeding your plant monthly, until autumn, with a balanced fertilizer.

If the air in your room is too dry for this plant, which flourishes in high humidity, sit its container on a tray of expanded clay granules and be sure to keep them moist.

As the never-never plant ages, it tends to lose its neat, compact shape. If this happens, remove any shoots that spoil the silhouette. Repot the plant every other year in multipurpose compost and if it becomes pot-bound divide the plant in spring to create the next generation.

PROBLEM SOLVING If never-never plants become too dry, their leaves will start to curl up. They'll soon suffer during the summer months if their compost has been allowed to dry out between watering. Avoid placing your plant in a cold draught, as this can cause it to shed leaves. Red spider mites can also become a problem if the air is too dry around the plant (see p.47).

DISPLAY

A shade-lover with a neat, compact shape when young, the never-never plant is perfect as a tabletop specimen or for positioning in front of a larger plant that's in an ugly container you may be keen to conceal. Alternatively, plant it under a larger specimen such as a kentia palm (see p.100) or a Swiss cheese plant (see p.109).

The never-never plant has a dense habit with a generous amount of foliage.

The herringbone plant is grown for its magnificently intricate foliage.

ALSO TRY

For similar striking foliage plants, look to relatives of the never-never plant such as goeppertias or marantas. All of them enjoy similar growing conditions to the never-never plant, including:

• **Herringbone plant** (*Maranta leuconeura* var. *leuconeura* 'Fascinator'), height 30cm (12in). Ideal for a hanging basket, the herringbone plant has spectacular light and dark green leaves with distinctive red leaf veins. It's an undemanding, easy-to-grow plant.

SAGO PALM *CYCAS REVOLUTA*

The sago palm is a cycad – an ancient, slow-growing plant – rather than a true palm, but its chunky trunk and arching, fern-like fronds certainly offer a tropical look. This is the perfect statement plant for a bright hallway and its architectural shape makes it a popular choice.

HEIGHT 90cm (36in)
SPREAD 90cm (36in)
FLOWERS White, feather-like
FOLIAGE Long, serrated
LIGHT Filtered sun
TEMPERATURE 13–24°C (55–75°F)
CARE Easy
PLACE OF ORIGIN China, southern Japan
WARNING! All parts are toxic; leaves are sharp; gloves required when handling

CARE

This slow-growing plant is often sold as a small specimen, but it will grow to be a fine, impressive plant in time. Its fronds are robust and prickly, so it needs space when mature. The sago palm is a terrific feature for a large entrance hall or reception area.

A room that has filtered light and moderate humidity is the ideal spot for this plant. To keep it healthy, allow the top of the compost to dry out between spring and summer and reduce watering further in winter.

Feed with a half-strength balanced fertilizer once a month in spring and summer, and grow in a 1:1 mix of multipurpose compost and loam-based compost. This is a top-heavy

> **DID YOU KNOW?** THE SAGO PALM IS AN ANCIENT PLANT THAT WAS DOMINANT DURING THE JURASSIC PERIOD, SO IT HAS BRUSHED SHOULDERS WITH THE DINOSAURS.

plant, but the weighty, loam-based compost will help it to remain upright. Repotting is only necessary every three years, which is good news if you're growing a large specimen.

Given the right conditions, this palm will be with you for life. When very mature, plants produce offsets that can be removed and potted on.

PROBLEM SOLVING Overwatering or watering into the crown of the plant can cause the sago palm to rot. If the lower leaves of the plant turn brown, simply cut them off – this is quite normal. When the tips of the leaves turn brown, increase the humidity around the plant by misting it with water. Red spider mites (see p.47) can be an issue for the sago palm.

DISPLAY

This is a plant that can provide a dramatic, eye-catching display all on its own, especially when planted in a stylish, generous container. To create a really tropical look, group it with some true palms such as the date palm (see p.121) and the bamboo palm (see p.125). Plant them all in individual containers so that you can turn the pots to prevent them from leaning towards the light.

Position this plant carefully to avoid injury from its needle-like leaves.

ALSO TRY

The stout, woody trunk of the sago palm is almost as fascinating as its dramatic foliage. If interesting trunks and stems appeal, then try growing this unusual plant alongside your sago palm – it will also thrive in filtered light and moderate humidity:

- **Money tree** (*Pachira aquatica*), height 1.8m (6ft). This large, distinctive plant is a mixture of a tree and a palm in appearance. It has elegant foliage and a striking, woody stem that's often braided. The money tree is said to bring good luck into the home and is frequently displayed in hallways.

Mature money trees are often sold with plaited stems. These plants have lush, glossy foliage that's divided into leaflets.

FLORIST CYCLAMEN

CYCLAMEN PERSICUM

This popular houseplant has huge appeal as a gift in winter. Florist cyclamen offers cheery blooms in the depths of winter, when days can be short and gloomy. It rarely lives after flowering, but its seasonal display as a living decoration, especially at Christmas, makes it well worth the investment.

HEIGHT 25cm (10in)
SPREAD 30cm (12in)
FLOWERS Pink, red, or white
FOLIAGE Green with silver markings, heart-shaped
LIGHT Filtered sun
TEMPERATURE 10-15°C (50–59°F)
CARE Fairly easy
PLACE OF ORIGIN Eastern Mediterranean

Florist cyclamen's blooms are available in red, white, and vibrant pink.

CARE

To ensure the much-admired flowers of the florist cyclamen don't fade too quickly, buy one in bud rather than in full bloom and place it on a north-facing windowsill in a cool room. Water sparingly, keeping the loam-based compost just moist. Feed fortnightly with a balanced liquid fertilizer.

These are short-lived plants that don't usually thrive for long after flowering (September to January), so repotting isn't necessary. It's not uncommon for gardeners to compost florist cyclamen after flowering as the foliage can quickly start to look unkempt.

If attempting to keep the plant for another year, reduce watering dramatically once the flowers fade and leaves start to yellow in late winter as this will encourage the tuberous plant to go dormant. Place the container in a cool room until July, when the plant should be repotted in a loam-based compost (see pp.30–31) – make sure its tuber is positioned just above the compost surface. Water the plant and position it somewhere that has filtered sun. Fresh foliage will form in summer, followed by flower buds in autumn.

PROBLEM SOLVING Yellow leaves are often the result of the cyclamen being placed in a room that's too hot and bright. Overwatering can cause grey mould on the leaves – pull off these leaves and reduce watering.

DISPLAY

Even before it flowers, this cyclamen will add interest to any indoor display, thanks to its attractive, heart-shaped leaves. More appealing for some is the prospect of a flowering houseplant during Christmas and you can add impact by displaying a group of them together. Combine your cyclamen with poinsettia (see p.85) and a scented, paper white narcissus (see p.110).

ALSO TRY

There are a number of excellent options for colourful houseplant displays over the festive season. Like the florist cyclamen, this hyacinth will flourish in a cool room that has filtered sunlight:
- **Dutch hyacinth** (*Hyacinthus*), height 20cm (8in). You can buy "prepared" bulbs in late September if you're hoping for flowers at Christmas time. The blooms of this plant are highly scented.

Choose large, firm Dutch hyacinth bulbs to produce the best flowers.

CYMBIDIUM *CYMBIDIUM*

One of the most widely grown orchids, cymbidium offers great variety. With numerous species and hybrids available, there's a tremendous choice of flower colour. Large orchids are ideal for a room that adjoins a patio, as these are plants that enjoy long, cool, summer nights.

HEIGHT 80cm (32in)
SPREAD 80cm (32in)
FLOWERS Stems of blooms
FOLIAGE Long, strappy
LIGHT Filtered sun
TEMPERATURE 10–24°C (50–75°F)
CARE Fairly easy
PLACE OF ORIGIN Asia and northern Australia

CARE

Cymbidiums are one of the easiest orchids to grow and their flowers are exceptional. Given the right conditions, the blooms will offer weeks of interest. To initiate the flowers requires cool nights in spring and summer and a room with moderate humidity and filtered sun, so avoid placing the plant in a south-facing room. If it has been kept cool enough at night, by autumn huge spikes of spectacular blooms will appear.

> **DID YOU KNOW?** YOU CAN BUY CYMBIDIUM BLOOMS AS CUT FLOWERS. THEY'RE COSTLY, BUT WILL LAST FOR A GOOD LENGTH OF TIME IN A VASE IN A COOL ROOM.

Cymbidium likes cool summer evenings, so you can move plants outside at night.

For best results, plant cymbidium in orchid compost and don't overpot it.

Plant in a specialist orchid compost and avoid overpotting (putting in too large a pot), which will discourage flowering. Water with rainwater, which has the correct pH for this orchid. Keep the compost moist, never wet. Unlike most orchids, cymbidiums won't need clear pots as their roots don't require light. Feed in spring and summer with a specialist orchid fertilizer, following the instructions on the bottle.

The generous spikes of flowers will need the support of a cane. Expect the blooms to grace your home for six to eight weeks. Once they've faded, cut off the spike at the base.

PROBLEM SOLVING Cymbidium is prone to mealybugs and red spider mites (see *pp.46–47*). If growing in a greenhouse, consider managing the mealybugs by biological control – that is, introducing a natural predator to attack and kill them off.

If the plant's leaf growth looks limp, this may be due to the cymbidium being placed somewhere too shady, so move your plant to a sunnier room.

DISPLAY

As cymbidium is such a large orchid, it can happily be displayed on the floor or on a tabletop. Take advantage of the fact that this orchid doesn't require a clear pot by choosing a container that will complement the plant's distinctive features and the style of your interior.

Orchids look stunning displayed in a group, so pair cymbidium with a moth orchid (see *p.118*), or some other variety. Alternatively, pair it with a contrasting plant, such as a begonia (see *p.59*), with its dramatic foliage.

ALSO TRY

Orchids are highly collectable, but not all enjoy similar conditions. However, this orchid has the same care needs as cymbidium and also doesn't require a clear pot:

• **Slipper orchid** (*Paphiopedilum*), height 30cm (12in). This strikingly beautiful orchid, named for its shoe shape, is easy to grow and comes in stunning colours.

UMBRELLA PALM *CYPERUS*

The umbrella palm is grown for its shape and its light, airy foliage, which resembles the spokes of an umbrella. This is an elegant plant that thrives in boggy, marshy soil in its native environment. If you're often guilty of watering plants too much, the umbrella palm is ideal for you.

HEIGHT 1.5m (5ft)
SPREAD 60cm (24in)
FLOWERS Insignificant
FOLIAGE Radiating, strap-like
LIGHT Filtered sun/light shade
TEMPERATURE 10–21°C (50–70°F)
CARE Fairly easy
PLACE OF ORIGIN Subtropical and South Africa

CARE

Some gardeners are tempted to keep their umbrella palm outside during the summer months, but this is a tender plant that needs the warmth of indoors to survive in autumn and winter. It will flourish in a moist environment in filtered sun or light shade, making it a fairly easy plant to place in the home.

The umbrella palm is classified as an aquatic plant and its roots can withstand being submerged in water – it's therefore virtually impossible to overwater. Plant the umbrella palm in multipurpose compost. Keep it soaked and place a dish full of water under the container. Mist the leaves regularly. Feed a couple of times during the spring and summer months with a balanced liquid plant food and prune off any dead stems to keep the plant in top condition.

PROBLEM SOLVING As the compost the umbrella plant is growing in is constantly moist, it's not unusual for the container to begin to smell. Resolve this issue by repotting in spring and adding charcoal to the compost.

Mealybugs (see *p.46*) can sometimes be a problem with this plant. Check regularly for the pests on the underside of the leaves or in the leaf joints. As this is a water-loving plant, you can try to remove the mealybugs by blasting the leaves of the plant with a jet of water – a task that's best done outside.

DISPLAY

The umbrella palm looks striking in a decorative pot with a matching saucer. It's a water-loving plant that will happily stand in water, so if you're planning an indoor water feature in your home, it makes a spectacular indoor display when lit up at night from below with underwater lights (see *p.42*).

This is not an easy plant to find a partner for as very few plants can cope with such moist compost. Mind-your-own-business (see *p.131*) is a rare exception: this low-growing plant offers a total contrast in shape and size to the umbrella plant but it also enjoys generous watering, making it an ideal companion.

The umbrella palm is classed as an aquatic plant and needs regular watering.

Monkey cups' pitchers hold nectar that insects are drawn to but drown in.

ALSO TRY

If you enjoy watering plants, it makes sense to focus on those that benefit from the attention. Carnivorous plants are thirsty in spring and summer: their roots must be moist, not soaked in water. Why not try:

- **Monkey cups** (*Nepenthes* hybrids), height 30cm (12in). Attractive, dark red pitchers dangle from these carnivores.
- **North American pitcher plants** (*Sarracenia* species and hybrids), height 30cm (12in). These plants have decorative and striking upright pitchers.

DUMB CANE *DIEFFENBACHIA*

With its huge, green and cream variegated leaves, dumb cane is a popular houseplant that adds sparkle and light to a gloomy corner – the perfect feature for an entrance hall. This is a plant that likes moderate humidity, so it's also a great choice for a kitchen.

HEIGHT 1.5m (5ft)
SPREAD 80cm (32in)
FLOWERS Insignificant
FOLIAGE Green and cream, variegated
LIGHT Filtered sun/light shade
TEMPERATURE 16–23°C (61–73°F)
CARE Fairly easy
PLACE OF ORIGIN Brazil
WARNING! All parts are toxic; gloves required when handling

CARE

A shady room in summer will be ideal for the dumb cane, but in winter move it into a brighter room. This impressive plant suffers if it's exposed to drafts or sudden changes in temperature, so be sure to choose its location carefully.

Plant in multipurpose compost. Dumb cane grows best in a humid environment, so place the container on a tray filled with soaked clay granules and mist the plant once a week. Keep the compost moist throughout the spring and summer months but only just moist during the winter. Feeding the plant once a month from spring to autumn will help accelerate growth.

DID YOU KNOW? IN COSTA RICA, POISON DART FROGS CLIMB UP GIANT DUMB CANE TREES AND DEPOSIT THEIR TADPOLES TO DEVELOP IN THE PLANT'S LEAF AXILS, WHERE RAIN COLLECTS.

Repot dumb cane every other year in spring, or more often if the plants become pot-bound.

As they age, dumb canes can start to look leggy. Resolve this by cutting the plant back to a stump of about 6cm (2½in) and wait for the new leaves to grow. All parts of this plant are toxic, so wear gloves when handling.

PROBLEM SOLVING If the base of the stem is discoloured, this is probably due to overwatering and too cool a temperature. Yellowing lower leaves often result from the plant being in a draughty spot. Dumb canes are more likely to suffer from a physical disorder than from pests, but red spider mite (see p.47) can be an issue.

DISPLAY

Dumb cane can look spectacular in a large container alongside a shade-loving prayer plant (see p.107) and a Boston fern (see p.112). Place the dumb cane at the back and the prayer plant and fern at the front of a display – the prayer plant trails while the fern has an arching habit. Display the plants in white, silver, or cream pots, which will pick up on the dumb cane's variegated foliage.

A young and healthy dumb cane will produce a neatly shaped plant.

ALSO TRY

Houseplants with patterned foliage are always a popular choice. They inject brightness to a dark room and offer interest that can be reflected in the style of the interior. For other plants with attractively marked and variegated foliage, why not try:

- **Angel wings** (*Caladium*), height 75cm (30in). The plant's arrow-shaped leaves have markings in green, white, pink, and red.
- **Zebra plant** (*Cryptanthus zonatus*), height 25cm (10in). This spidery-looking bromeliad has funky, burgundy foliage with cream stripes. It likes heat and light.

The highly decorative, bicoloured leaves of *Caladium* 'White Christmas' (left), and 'Rosebud' are shown here.

VENUS FLYTRAP *DIONAEA MUSCIPULA*

The Venus flytrap – famous for its snapping, jaw-like leaves – is a source of fascination to children and adults alike. It needs direct sun, so a south-facing room is ideal. This popular carnivorous plant really does catch flies, which also makes it a terrific plant for the kitchen.

HEIGHT 10cm (4in)
SPREAD 20cm (8in)
FLOWERS White, tubular
FOLIAGE Jaw-like
LIGHT Sun
TEMPERATURE 9–27°C (48–81°F)
CARE Fairly easy
PLACE OF ORIGIN Southeastern USA

The Venus flytrap is the perfect pocket-money plant for children who have a sunny windowsill in their bedroom.

CARE

Venus flytrap flourishes in a humid environment with plenty of sunlight. A sunny, eye-level windowsill, where the plant's snapping leaves can be easily observed, is the perfect location.

Choose a small container with a saucer, and plant the Venus flytrap in a specialist carnivorous plant compost. During the spring and summer months, keep the saucer under the container full of water and the compost moist. In winter, remove the saucer, keep the compost moist, and move the plant to a slightly cooler room. Watering with rainwater is essential as the calcium that's present in some tap water will eventually kill this plant. Collecting rainwater is always a major part of the fun for young gardeners.

You won't need to feed the Venus flytrap as it lives off flies, but you will need to open the window occasionally to allow insects to enter. If flowers arrive in springtime, simply cut them off as they'll exhaust the plant.

PROBLEM SOLVING It's more likely that your flytrap will eat pests rather than be attacked by them. However, mealybugs (see p.46) can be an issue, so watch out for the white insects in the crown of the plant. In the majority of cases, these pests can be wiped off quite easily with a cloth.

DID YOU KNOW? WHEN A FLY ENTERS A LEAF TRAP IT SNAPS SHUT AROUND THE INSECT. IT TAKES ABOUT A WEEK FOR THE PLANT TO DIGEST THE FLY.

DISPLAY

As the Venus flytrap is such a small plant, it's often grown in a large container with other carnivorous plants. An indoor window box is also a great place to grow a group of these plants as it will neatly fit into a sunny, indoor windowsill. The Cape sundew (see p.82) – a carnivore of a similar scale to the Venus flytrap – makes a good partner plant.

ALSO TRY

Insects are attracted to carnivorous plants by the scent of their nectar or their colourful pitchers (modified leaves). The insects get trapped in the pitchers or stuck on the leaves. All carnivorous plants enjoy the same growing conditions, so it's easy to start a collection. Why not try:
- **California pitcher plant** (*Darlingtonia californica*), height 40cm (16in). The red-veined pitchers of this plant exude a honey scent, which attracts flies.
- **Pitcher plant** (*Sarracenia purpurea*), height 15cm (6in). This low-growing plant has short, deep, burgundy-coloured pitchers.

DRAGON PLANT *DRACAENA FRAGRANS*

When mature, this variegated plant has a thick, woody-looking stem and will produce multiple crowns. Most dragon plants have green leaf edges with yellow central stripes, but there are variations. The glossy leaves add sparkle to a shady room, which make this easy-to-grow plant a popular choice.

HEIGHT 1.2m (4ft)
SPREAD 90cm (36in)
FLOWERS Small, white
FOLIAGE Strappy, variegated
LIGHT Filtered sun/light shade
TEMPERATURE 15–24°C (59–75°F)
CARE Easy
PLACE OF ORIGIN Tropical Africa

CARE

Dragon plants are straightforward to place in the home as they thrive in both filtered sun and light shade. They require moderate humidity so will be equally happy in a sitting room or a bathroom.

The plant should be kept moist but not wet in spring and summer. Wipe the leaves regularly with a damp cloth to keep them shiny and healthy. During this time, feed with a balanced fertilizer every two weeks and mist occasionally with water. Reduce watering in the winter and keep the plant away from cold draughts.

Larger specimens can be top-heavy, so choose a peat-free, loam-based compost. This will help the plant stay upright as it's heavier and offers more support to the roots than multipurpose compost. If the plant has grown taller than you'd like, cut the stem down to a preferred height and it will resprout.

PROBLEM SOLVING The dragon plant is relatively pest-free. The most common ailment is brown tips on the leaves, which is frequently caused by cold draughts or dry air. Using a pair of florist scissors (see *p.32*), simply cut off the brown tips, making sure that the leaves are left with a natural shape.

DISPLAY

Smaller specimens are perfect for a tabletop or mixed bowl of houseplants, while mature plants offer a tropical look to a room and add height. Mature plants with trunks and several crowns are excellent specimens for a decorative pot, but make sure it's sturdy enough to support the plant.

Dragon plants are easy to care for, so they'll combine well with other shade-tolerant plants. For an eye-catching mixed container, partner them with plants that have contrasting foliage, such as the radiator plant (see *p.117*) and the peace lily (see *p.133*).

ALSO TRY

If you're hoping to grow a plant to a specific height – as you can with the dragon plant – there are several climbing houseplants that can be trained up moss poles, including:

• **Blushing philodendron** (*Philodendron erubescens* 'Red Emerald'), height 1.2m (4ft). The stunning red stalks and large, heart-shaped leaves of this plant will happily climb a moss pole.

• **Jungle vine** (*Parthenocissus amazonica*), height/trail up to 1.2m (4ft). This plant has arrow-shaped leaves with attractive silver veins and maroon undersides.

A young dragon plant makes a bold statement in a modern setting.

Keep leaves shiny and healthy by cleaning with a damp cloth.

MADAGASCAR DRAGON PLANT *DRACAENA MARGINATA*

HEIGHT 1.5m (5ft)
SPREAD 90cm (36in)
FLOWERS Small, white
FOLIAGE Strappy, variegated
LIGHT Filtered sun/light shade
TEMPERATURE 15–24°C (59–75°F)
CARE Easy
PLACE OF ORIGIN Madagascar

The undemanding Madagascar dragon plant is grown for its palm-like appearance. It's the perfect specimen plant for a shady spot, but its green, red-edged, sword-like leaves make it an attractive and popular addition to any room.

The Madagascar dragon plant is one of the easiest houseplants to grow.

CARE

The Madagascar dragon plant is easy to place as it can thrive in filtered light or shade and requires only moderate humidity. Being so simple to care for, it's a popular choice for workspaces and offices as well as for the home.

The roots are relatively small, so choose a pot that fits them snugly and avoid overpotting. Taller specimens can topple – avoid this by using loam-based compost, which is heavier than other types of compost and will give stability.

Mist your plant once a week if it's placed in a room that has low humidity.

In spring and summer, feed with a half-strength balanced fertilizer every two weeks and keep plants moist but never wet. In winter, reduce watering so that the compost remains just moist.

Keep your plant in top condition and looking great by misting occasionally, removing untidy lower leaves, and wiping the foliage with a damp cloth. If you wish to propagate your own plants, take cuttings in spring.

PROBLEM SOLVING Dracaenas are more likely to suffer leaf damage due to their environment rather than to pests. They won't tolerate draughts, sudden changes in temperature, or overwatering in winter – this will cause their leaves to fall or go brown. Any damaged leaves should be removed, or the tips cut off with florist scissors.

DISPLAY

With its wonderfully stately habit, the Madagascar dragon plant is perfect as the statement plant in a large container of mixed houseplants. To make the most of this plant's flamboyant, spiky foliage, grow it alongside plants that have contrasting foliage shapes. Ideal partners, which will also cope with filtered light or shade and moderate humidity, include the Chinese evergreen (*see p.52*) and the equally unfussy dumb cane (*see p.77*). All these plants have variegated foliage, but in markedly different styles. Together, this trio will add life, interest, and vibrancy to a gloomy corner of the home.

ALSO TRY

The Madagascar dragon plant's maroon-edged foliage is very attractive, so consider linking the plant to a room based on warm colours, including the following plant:
- **Three-coloured Madagascar dragon tree** (*Dracaena marginata* 'Tricolor'), height 1.5m (5ft). This plant is identical to *Dracaena marginata*, but has foliage with cream, red, and green stripes. To keep variegation strong, it needs a sunnier room than its relative.

LUCKY BAMBOO

DRACAENA SANDERIANA

This plant takes its name from the traditional Chinese belief in bamboo as a symbol of strength and good fortune, which has made it popular as a house-warming gift. Its stems resemble bamboo and grow straight, but the plant is often sold with twisted stems, trained by the grower.

HEIGHT 90cm (36in)
SPREAD 20cm (8in)
FLOWERS Insignificant
FOLIAGE Bright green
LIGHT Filtered sun
TEMPERATURE 18–27°C (64–81°F)
CARE Easy
PLACE OF ORIGIN Tropical West Africa

CARE

Choose a bright room with moderate humidity for the lucky bamboo and place it so that it can be easily turned – this stops the stems growing towards the light. If you've struggled with plants due to low levels of natural light, the lucky bamboo may be the answer as it thrives under fluorescent lighting.

You can very easily grow this plant in water. To do this, fill a vase with pebbles (to help hold stems upright) and distilled water or rainwater, but not tap water. Change the water weekly. Alternatively, plant in multipurpose compost, keeping the compost moist. Feed twice from spring into summer with a balanced fertilizer; feed vase-grown plants in exactly the same way.

The twisted stems of lucky bamboo are easy to grow in compost or water.

DID YOU KNOW? IN ITS NATIVE HOME OF TROPICAL WEST AFRICA, THE LUCKY BAMBOO GROWS TO AN IMPRESSIVE 1.5M (5FT) IN HEIGHT AND SPREAD. ALTHOUGH IT CLOSELY RESEMBLES A TRUE BAMBOO IN BOTH SIZE AND APPEARANCE, IN FACT IT'S COMPLETELY UNRELATED.

Keep the lucky bamboo in good shape by pruning the shoots to about 5cm (2in) from the main stem. Place the pieces you trim off in jars of water to root and, hopefully, make new plants.

PROBLEM SOLVING Yellow leaves are often the result of too much bright sunlight or overfeeding with fertilizer. When growing in water, if a stem goes soft at the end, cut the stem down to a healthy and firm piece of plant material and place in a vase of water. If you grow the lucky bamboo in tap water, the leaves are likely to turn brown.

DISPLAY

Twisted, plaited, or bunched stems in a clear vase can make a terrific centrepiece for a coffee table. Alternatively, plant lucky bamboo in compost as part of a group display, alongside a small money plant (see p.71) and the zebra cactus (see p.96). Both of these plants need low humidity and will make good partners for lucky bamboo.

Stems of lucky bamboo will grow happily in distilled water or rainwater.

ALSO TRY

If you like the fresh, modern look of living plants growing in vases or jars of water, you could also try:

- **Satin potho** (*Scindapsus pictus* 'Argyraeus'), height/trail 1m (3ft 3in). The heart-shaped leaves of this trailing plant have stunning, silver markings.
- **Small-leaf spiderwort** (*Tradescantia fluminensis*), height/trail 60cm (24in). This is an attractive, easy-to-grow, white- and green-leaved trailing plant.

CAPE SUNDEW *DROSERA CAPENSIS*

In its natural habitat, the Cape sundew enjoys an open site with moist soil, where it receives plenty of sunlight; indoors, the small plant will thrive on a sunny windowsill or beneath fluorescent lighting. The plant's sticky leaves, which ensnare insects, are by no means dramatic, but they are intriguing.

HEIGHT 15cm (6in)
SPREAD 20cm (8in)
FLOWERS Insignificant
FOLIAGE Long and slim with tentacles
LIGHT Sun/filtered sun
TEMPERATURE 7–29°C (45–84°F)
CARE Fairly easy
PLACE OF ORIGIN Cape of South Africa

CARE

If you're hoping to encourage your children to care for houseplants, this is a great way to start as the Cape sundew is a relatively easy carnivorous plant to maintain.

The long, slim leaves of the small yet fascinating plant are covered in colourful tentacles that exude a sticky substance called mucilage. Unsuspecting insects get stuck on the leaves, which then quickly curl around the prey. Over time, the plant will absorb the insect, so it won't require any additional plant feed. The Cape sundew only needs to ensnare about three flies a month in order to remain healthy.

Cape sundews should be planted in specialist carnivorous plant compost, which is widely available. Keep the roots moist and place the container on a saucer that's full of rainwater (tap water will harm this plant). In its natural habitat, the sundew will go dormant in winter, but in a warm home it should remain actively growing so will require fairly generous watering all year round. Remove any flowers that appear as these will only weaken the plant.

PROBLEM SOLVING If Cape sundews don't receive the correct amounts of light, warmth, humidity, or fresh air, they commonly fail to produce the sticky mucilage. If this happens, assess your room and try to work out which of the plant's essential requirements might need adjusting.

Over-enthusiastic watering can be an issue – leaves will fall off or rot. Plants must be moist, not soaking wet.

DISPLAY

Sundews are often displayed with other small plants in a child's bedroom. A room with a south-facing windowsill is ideal. In a sunny spot, living stone (see p.106) and bunny ears cactus (see p.115) will live happily alongside the sundew in separate containers, but they'll require very little watering compared to the carnivorous plant.

The plant's tentacles are covered in a sticky substance that captures insects.

Butterwort uses its sticky leaves to ensnare small insects.

ALSO TRY

With so many different and fascinating carnivorous plants available from specialist growers, you can soon become a collector. Buy a large drip tray that will fit your windowsill and place pots of other meat-eaters on it, such as:

- **Butterwort** (*Pinguicula*), height 15cm (6in). Like the Cape sundew, this is a delicate-looking but deadly plant.
- **North American pitcher plants** (*Sarracenia* species and hybrids), height 30cm (12in). This striking plant has large, beautifully patterned pitchers.

ECHEVERIA *ECHEVERIA*

Echeverias are neat, compact plants formed by tight rosettes of spoon-shaped leaves in blue-green, red, purple, or silver. These succulents are easy to grow and adore a south-facing room or sunny windowsill. If it's warm enough, you'll be blessed with stems of pink and yellow lantern flowers in late summer.

HEIGHT 10cm (4in)
SPREAD 30cm (12in)
FLOWERS Stems of pink or yellow blooms
FOLIAGE Succulent rosettes
LIGHT Sun/filtered sun
TEMPERATURE 10–30°C (50–86°F)
CARE Easy
PLACE OF ORIGIN Central America and Mexico

Choose round- and pointed-leaved echeverias to create a contrasting display.

CARE

Plant echeverias in individual pots or put several species in a larger dish and position them in a bright spot with low humidity. Place them outside in summer but bring them inside for protection during the autumn and winter.

These plants are natives of arid environments, so place them in a cactus compost and keep the watering to an absolute minimum – echeverias are perfect for people who are often away from home. Avoid watering the crown of the plant as this can lead to rot. Apply a half-strength liquid feed every couple of weeks during the summer months.

Echeverias are easy to propagate: simply cut off the offsets in spring and give them their own pot. You can also propagate by taking healthy leaf cuttings and pushing them into moist compost.

Repot echeverias every other year. This can be a fiddly task as the leaves have a tendency to fall off.

PROBLEM SOLVING The most persistent and widespread problems with echeverias are caused by either too much or too little watering. Both these issues cause the plant to wilt, turn yellow, and then shrivel. Plants should dry out between watering; after a good water, put them on a draining board for a while. If light levels are too low, plants can become leggy. Remedy this by moving them to a brighter spot.

DISPLAY

These architectural plants are a popular choice for anyone wanting to create a striking display in a small, contemporary home. You can have great fun displaying your echeverias in different ways: treat them to interesting individual pots; plant several together in a large, shallow pot, or line a basket with plastic and use this as a container. Echeverias look great displayed alongside other succulents such as the money plant (see *p.71*).

ALSO TRY

Many indoor gardeners choose to fill their sunny home with easy-care succulents. They offer a modern look and don't require much attention. For distinctive, architectural shapes, also try:

- **Donkey's tail** (*Sedum morganianum*), height 10cm (4in). Strands of trailing stems covered in succulent leaves make this plant the ideal choice for a hanging basket or a high shelf.
- **Pincushion euphorbia** (*Euphorbia enopla*), height 30cm (12in). Despite this plant's bright red, cactus-like prickles, it isn't a cactus. It's perfect for a high shelf.

Grow donkey's tail in cactus compost to achieve the best results.

DEVIL'S IVY *EPIPREMNUM AUREUM*

There aren't many climbing or trailing houseplants as reliable as this one. Devil's ivy will happily trail from a hanging basket, climb a trellis, or leap up a moss pole. The striking, versatile plant is widely grown for its arrow-shaped leaves, with their attractive marbled, yellow markings.

HEIGHT 2m (6ft 6in)
SPREAD 2m (6ft 6in)
FLOWERS Unlikely to bloom indoors
FOLIAGE Heart-shaped, variegated
LIGHT Filtered sun/shade
TEMPERATURE 15–24°C (59–75°F)
CARE Easy
PLACE OF ORIGIN Indonesia

CARE

Devil's ivy is a foliage plant with aerial roots. It's perfect for beginners as it will survive in any room, apart from those with intense, direct sunlight. In its native jungle habitat it's quite an invasive plant, but restricted to a container it's more manageable. Plant in a multipurpose compost on its own or with a selection of other houseplants. Water well in spring and summer but allow plants to dry out a little between watering. In winter, keep it just moist. It will grow without a plant feed, but ensure it's healthy by feeding monthly in spring and summer with a balanced fertilizer. Clean the leaves with a damp cloth.

Use ties or clips to train the plant up a moss pole or trellis. Keep devil's ivy in shape by pruning excess growth in the springtime. You can use some of the pruned-off stems as cuttings and root them in a jar of water.

PROBLEM SOLVING Yellowing and falling leaves are most likely to be the result of overwatering or of the plant being subjected to cold draughts. Marked or damaged leaves can simply be removed. If the plant's leaf variegation fades, this is probably due to a lack of light. This robust plant rarely suffers from pests and diseases.

DISPLAY

Decide if you want to grow devil's ivy as a trailing or climbing specimen – it will happily perform either role. This is a versatile plant that can be used to add height to a room, cover a room-divide trellis, or offer a wall of foliage when allowed to trail from a balcony. It can also be used at the front of a mixed display to cover an ugly pot. If you'd like a collection of hanging baskets in your home, then grow it alongside other trailing plants such as grape ivy (see *p.66*) or creeping fig (see *p.90*).

When devil's ivy reaches maturity its variegation may weaken, but it's still an immensely impressive plant.

ALSO TRY

Using a moss pole to train a trailing plant to grow up rather than down is easy. Other plants that work well grown in this way include:

- **Ivy** (*Hedera*), height varies. Ivy is perceived as an outdoor plant, but if you have a cool room, it can look stunning indoors climbing up poles or tumbling from baskets.
- **Variegated Swiss cheese plant** (*Monstera deliciosa* 'Variegata'), height 2.5m (8ft). This striking plant has giant white and green variegated leaves.

Monsteras have long aerial roots that shouldn't be cut off.

POINSETTIA *EUPHORBIA PULCHERRIMA*

Christmas wouldn't be complete without a poinsettia in the house – they're a symbol of the festive season. Many people throw away their poinsettias as soon as their colour fades, but more adventurous indoor gardeners, keen on a challenge, endeavour to keep them going for longer.

HEIGHT 60cm (24in)
SPREAD 30cm (12in)
FLOWERS Yellow with red bracts
FOLIAGE Dark green
LIGHT Sun/filtered sun
TEMPERATURE 13–27°C (55–81°F)
CARE Challenging
PLACE OF ORIGIN Central America and Mexico
WARNING! Sap is toxic; gloves required

The poinsettia is often bought as a temporary living decoration.

CARE

Poinsettias are sold in the thousands at Christmas as gifts or living decorations. Although it's tempting to place them on the mantelpiece by a winter fire, they won't survive close to excessive heat. They are far better placed on the centre of a dining table as they require filtered light and no draughts.

There's rarely a need to repot these plants if you're keeping them for just one season. Overwatering can cause problems with poinsettias; water cautiously, allowing the plant to begin to dry out before watering again.

When growing for one season only, you won't need to worry about feeding. Mist your plants once a week.

Once the poinsettia leaves have fallen, and if you want to keep the plant for another year, prune it back to 10cm (4in) in April and keep it in a cool room, 13°C (55°F). If repotting is required, use multipurpose compost. To encourage bracts, from November, plants should receive no more than 12 hours of natural or artificial light daily and a temperature of 18°C (64°F).

PROBLEM SOLVING It's not uncommon for a poinsettia to start to wilt as soon as you get it home. To avoid this, ask the seller to wrap it with paper all the way around the top. A poinsettia that has undergone cold conditions at the shop or on the journey home can die (see p.27). Even a short spell in cold outdoor temperatures can prove fatal.

DISPLAY

Modern varieties of poinsettia are compact and some forms have red, white, or pink bracts that can keep their colour for months. As it's such a flamboyant plant, it rarely needs a partner. Choose a seasonal container or simply wrap a big bow around the pot and sit it on a saucer to protect your table. If you want to create a festive floral scene for your guests, display your plant with a cyclamen (see p.74) and a bowl of scented paper whites (see p.110).

ALSO TRY

Short-lived plants that are bought for immediate, temporary impact have great value. If you'd like to introduce more colour and impact into your home after the poinsettia's flowers fade, consider these plants:

- **Cineraria** (*Pericallis* × *hybrida*), height 30cm (12in). This plant is covered in purple, pink, or blue flowers. It's usually sold between December and May and is ideal for a room that has filtered light.
- **Lily of the valley** (*Convallaria majalis*), height 20cm (8in). If placed in a cool room, this pretty plant offers short-lived, scented white flowers. All parts of lily of the valley are toxic.

Cineraria is a flowering houseplant for a room with a cottagey feel.

CASTER OIL PLANT

FATSIA JAPONICA

This large, accommodating specimen plant is perfect for a room that has light shade and moderate humidity. With dark green, deeply lobed, glossy leaves that cover the entire stem, the castor oil plant is impressive from top to bottom and makes a striking addition to almost any style of interior.

HEIGHT 2m (6ft 6in)
SPREAD 2m (6ft 6in)
FLOWERS Insignificant
FOLIAGE Large, hand-shaped
LIGHT Filtered light/light shade
TEMPERATURE 10–25°C (50–77°F)
CARE Easy
PLACE OF ORIGIN Japan

This easy houseplant can also be grown outside in a sheltered garden.

CARE

This relatively undemanding plant is perfect for a beginner. If you're on a budget, invest in a smaller plant as this is a fast-growing genus. Enjoying light shade or filtered sunlight, it's easy to position in most homes, but will make a statement in a hallway or sitting room, thanks to its large, leathery leaves. Keep it away from draughts and direct heat and it will be with you for years.

The castor oil plant will survive in a multipurpose compost, but larger specimens are best in a heavier, loam-based compost, which will offer more support. In spring and summer, keep the plant moist, but reduce watering slightly in winter. Also during this time, feed the castor oil plant with a balanced fertilizer once a fortnight.

The impressive leaves shine easily when wiped over with a damp cloth, so this is worth doing regularly. To keep your plant bushy, lightly prune it to shape in the springtime and repot it in a generous container every year. Without this annual pruning the plant will become leggy and unattractive.

PROBLEM SOLVING The castor oil plant could be prone to an attack from any of the common houseplant pests but, thanks to its large leaves, they are extremely easy to spot. Another frequent problem is leaves that start to shrivel. This is most likely to result from air that's too dry. Resolve this by misting the plants regularly.

DISPLAY

Mature specimens look spectacular when grown in a large container and displayed on their own. Smaller plants can be used as part of a mixed planting and grown with other accommodating, undemanding plants such as the never-never plant (see p.72) and the Chinese evergreen (see p.52). You can pick up on the red undersides of the leaves of the never-never plant by planting your group in a red planter.

(see p.72) ... (see p.52)

ALSO TRY

The giant, sturdy leaves of the castor oil plant can't fail to add an eye-catching, tropical look to your home. If this kind of thick, lobed foliage appeals to you, consider growing this plant:

• **Variegated fatsia** (*Fatsia japonica* 'Variegata'), height 2m (6ft 6in). The growing habit and requirements of this attractive fatsia are almost identical to those of the caster oil plant. The thick leathery leaves of the variegated fatsia are narrowly edged with cream. To keep the variegation strong, it prefers more sunlight than its relative. Many gardeners with sheltered gardens grow this plant outside.

To keep small and compact, prune variegated fatsias regularly.

WEEPING FIG *FICUS BENJAMINA*

This is a hugely popular plant thanks to its graceful, arching habit and tree-like form. It can be bought as a small, tabletop plant or a large specimen. When mature, it makes the perfect centrepiece and is often positioned in a large hallway, where it can be enjoyed from all angles.

HEIGHT 3.5m (11ft 6in)
SPREAD 1.2m (4ft)
FLOWERS Unlikely to bloom indoors
FOLIAGE Small, dark green
LIGHT Filtered sun/light shade
TEMPERATURE 16–24°C (61–75°F)
CARE Challenging
PLACE OF ORIGIN Northern Australia, South and Southeast Asia
WARNING! All parts toxic; gloves required

CARE

Although this attractive plant is highly desirable, it isn't easy to grow. You'd be wise to choose a different plant if you're unable to meet its exact requirements. The weeping fig needs space, filtered sun or light shade, moderate humidity, and no draughts. Deviate from this and the leaves on your plant will quickly drop. However,

Weeping figs are available to buy at any size – from tabletop to head height.

those that have the perfect room won't regret the investment as the striking, dark foliage complements any interior.

Choose a large container and use a loam-based compost that will help hold the top-heavy plant steady. Allow the top of the compost to dry out between watering in spring and summer and reduce watering further in winter. A misting with water in summer is beneficial, as is applying a feed of balanced fertilizer once a month in spring through to autumn.

Repotting large specimens is tricky and can cause leaf drop, so if you're concerned, simply add a fresh layer of compost to the container each spring.

PROBLEM SOLVING The most likely cause of concern is leaf drop. This is due to stress caused by under- or overwatering, low humidity, not enough light, draughts, or a sudden change in temperature – you'll need to do some detective work to discover which one it is. If your plant has lots of yellow leaves, lightly shake it to remove them.

DISPLAY

Larger specimens look at their best when planted in a generous container and placed in a position where they have space all around them – don't push these plants into a corner. If growing in a large pot, you could add trailing plants around the bottom: devil's ivy (see *p.84*) or its much smaller relative creeping fig (see *p.90*) would work well.

ALSO TRY

If you've had success with weeping fig, there are many other members of this family to grow that are equally appealing. Why not try:

- **Bengal fig** (*Ficus benghalensis*), height 3.5m (11ft 6in). Although rarely grown, this plant looks terrific in a modern interior, thanks to its tree-like form and large leaves.
- **Variegated weeping fig** (*Ficus benjamina* 'Variegata'), height 3m (10ft). This is an identical plant to the plain weeping fig but with stunning variegation. Cut off any stems that revert to plain green.

If placing ficus on a windowsill, make sure it's not a draughty one.

RUBBER PLANT *FICUS ELASTICA*

The rubber plant is grown for its broad, glossy, dark green leaves and attractive, tree-like shape. It's a popular plant that's easy to please and copes well in a shady spot if protected from draughts. Enjoying low to moderate humidity, it makes the ideal specimen for a sitting room.

HEIGHT 1.8m (6ft)
SPREAD 1.2m (4ft)
FLOWERS Rarely blooms indoors
FOLIAGE Large, rubbery
LIGHT Filtered sun/light shade
TEMPERATURE 15–24°C (59–75°F)
CARE Easy
PLACE OF ORIGIN South Asia
WARNING! Sap is toxic; gloves required when handling

CARE

Unlike other species of *Ficus*, this accommodating plant isn't fussy. It copes well with low light levels and is also fairly drought tolerant, so it's a popular choice for forgetful gardeners.

If you buy a small plant it will soon grow into an impressive specimen, which makes it a sensible option for anyone who's looking for larger plants for their home on a budget. Choose a generous pot and plant in a loam-based

The rubber plant is eye-catching, fast-growing, and easy to maintain.

Clean the leaves with a damp cloth to keep them healthy and shiny.

compost with a few handfuls of perlite mixed in to assist with drainage. Rubber plants prefer a generous watering and then the chance to dry out before watering again. Make sure the compost is not consistently damp and never allow your rubber plant to sit in water. Feed fortnightly in spring and summer with a half-strength balanced fertilizer.

Rubber plants are dust magnets, so wipe the leaves with a damp cloth and it will look, and be, more healthy.

If kept for some time in a shady room, the rubber plant might become a bit leggy. Prune the top growth at least once a year in spring, which will give it a more balanced shape.

PROBLEM SOLVING If the leaves fall, the most likely cause is overwatering or draughts – both are easily solved. This plant isn't a stranger to attack

by mealybugs and scale insects (see pp.46–47). You may be able to remedy by wiping them away with a damp cloth.

DISPLAY

No matter how high the ceilings in your home, this plant will grow upwards to meet them. Most indoor gardeners keep their rubber plants well pruned so that they don't become too large. When smaller, they look striking with variegated plants such as the dumb cane (see p.77) and the variegated snake plant (see p.128). However, avoid planting them together in one pot as they have different watering requirements.

ALSO TRY

Growing a plant with such shiny leaves is very appealing to many indoor gardeners. If you're looking for other specimens with leaves to polish that will add gloss to your home, also try:

- **Flaming sword** (*Vriesea splendens*), height 60cm (24in). The foliage of this bromeliad has dazzling, lime-green stripes.
- **Rubber plant 'Tricolor'** (*Ficus elastica* 'Tricolor'), height 1.8m (6ft). This plant is identical to *Ficus elastica* but has pink-, white-, and green-patterned leaves and requires more sunlight.

FIDDLE-LEAF FIG *FICUS LYRATA*

This distinctive tree-like plant is named after its attractive, violin-shaped leaves. Its native home is a tropical rainforest, making it an appealing option for anyone keen to recreate a jungly look. The fiddle-leaf fig likes moderate humidity and filtered sun and is popular as a sitting-room plant.

HEIGHT 1.8m (6ft)
SPREAD 1.2m (4ft)
FLOWERS Won't bloom indoors
FOLIAGE Large, puckered
LIGHT Filtered sun
TEMPERATURE 15–24°C (59–75°F)
CARE Easy
PLACE OF ORIGIN West Africa
WARNING! Sap is toxic and stains; gloves required when handling

The leaves of the fiddle-leaf fig are similar to those of the edible garden fig.

CARE

When positioning the fiddle-lead fig, choose a spot with filtered light, where the container can be easily turned to keep the plant growing upright.

Select a generous container and plant in a non-peat, loam-based compost with a few handfuls of perlite mixed in to assist with drainage. Give the plant a generous watering and then a chance to dry out before watering it again – overwatering will lead to leaf drop. Feed every other week with a half-strength, balanced fertilizer during spring and summer.

If left to its own devices, this striking plant will grow to ceiling height, but it can quite easily be pruned back to a suitable size in spring. The sap can cause irritation to the skin, so wear gloves when handling and also try to prevent it from getting on carpets and clothing as it can stain.

PROBLEM SOLVING Sudden leaf drop by the fiddle-leaf fig is most likely to be due to draughts, a dramatic change in temperature, or overwatering.

This easy-to-grow plant will attract houseplant pests such as mealybugs and scale insects (see *pp.46–47*). If you act quickly, you might be able to simply wipe them off with a damp cloth.

DISPLAY

The fiddle-leaf fig tends to grow with one main stem, which gives it the much-desired look of a tree. For this reason, it's commonly sold as a mature plant that will make an instant statement in a room. If you have the space, give this fig the opportunity to stretch its limbs and grow into a large specimen. A good partner plant that enjoys the same conditions and will also grow to fit the room is the rubber plant (see *opposite*). Add a little more rainforest drama to the display with the rapid-growing Swiss cheese plant (see *p.109*).

ALSO TRY

When grouping houseplants together the shape, texture, and colour of the leaves should be considered. The fiddle-leaf fig's foliage is often used to add much-need texture to a grouping. Other plants with this quality include:

- **Donkey's tail** (*Sedum morganianum*), height 10cm (4in). This succulent trailer has rope-like stems covered in tiny, round leaves.
- **Spear sansevieria** (*Sansevieria cylindrica*), height 75cm (30in). The impressive, cylindrical leaves of this succulent remain bolt upright.

Donkey's tail is ideal for a south-facing room and requires little care.

CREEPING FIG *FICUS PUMILA*

The creeping fig likes filtered light and low humidity, so although it may be tempting to grow it in a hanging basket in the bathroom, it's better off in a sitting room or bedroom. This tiny plant is often grown beneath larger specimens. Its trailing stems can be trained to climb up a moss pole.

HEIGHT 90cm (36in)
SPREAD 90cm (36in)
FLOWERS Insignificant
FOLIAGE Small, green
LIGHT Filtered sun/light shade
TEMPERATURE 13–24°C (55–75°F)
CARE Fairly easy
PLACE OF ORIGIN East Asia

This versatile plant will tumble from a basket or climb a trellis or moss pole.

CARE

This versatile plant is ideal for a tight space. Place it in a small pot on your desk or windowsill, or on a bookshelf.

Plant creeping fig in multipurpose compost and apply a balanced liquid fertilizer once a month during the spring and summer. The key to real success is to keep the compost just moist – don't allow the plant to dry out or the leaves will soon shrivel and look unattractive. Don't leave the plant sitting in water as this will almost certainly lead to its death. Reduce watering slightly during the winter months.

Prune creeping fig in spring to keep the plant in shape and to your preferred size. At the same time, consider taking cuttings, which will root very easily. If you plan to grow the plant up a support, you'll first need to train it by guiding the stems.

PROBLEM SOLVING Very old plants will start to look scruffy, but pruning hard in spring will give them a new lease of life.

The creeping fig tends to remain pest- and disease-free, but can be affected by mealybugs and scale insects (see pp.46–47) if grouped with too many plants in a hot environment.

DISPLAY

Creeping fig is ideal for indoor hanging baskets. You can also encourage it to climb up a small, decorative plant support or a moss pole. This may not be the most flashy or dramatic plant, but it's extremely handy, and a great partner to more decorative plants. Consider planting creeping fig alongside other trailing plants to create a really striking mixed houseplant basket. The spider plant (see p.65) and waxflower (see p.101) will enjoy a similar position in the home and require a comparable watering regime.

ALSO TRY

Ficus pumila isn't the only versatile, trailing and climbing member of the *Ficus* group. Consider adding impact and interest to your display by introducing another fig to your indoor hanging garden. You could be really adventurous and grow the plants together in one container or train the stems to climb up a moss pole. Why not try:

• **Variegated creeping fig** (*Ficus pumila* 'Variegata'), height 90cm (36in). This plant is virtually identical to *Ficus pumila* but offers attractive, variegated foliage. If its leaves revert to green, cut off the whole stem to prevent the entire plant from turning green.

Variegated creeping fig can either be pruned back to keep it neat or left so that it can trail.

MOSAIC PLANT

FITTONIA ALBIVENIS VERSCHAFFELTII GROUP

This is a great pocket-money plant – small enough for any room and exciting enough for every interior. The neat, low-growing creeper has gloriously patterned foliage with pink or white veins. It enjoys high humidity and is a perfect addition to a terrarium, bottle garden, or bathroom display.

HEIGHT 15cm (6in)
SPREAD 20cm (8in)
FLOWERS Insignificant
FOLIAGE Patterned
LIGHT Filtered sun
TEMPERATURE 17–26°C (63–79°F)
CARE Fairly easy
PLACE OF ORIGIN South America

This low-growing, creeping plant offers a neat and tidy shape.

CARE

Mosaic plants will thrive in a room with filtered sun and, as moist air is vital to them, their ideal location is a terrarium.

Plant in multipurpose compost and keep moist in spring and summer, but reduce watering slightly in winter. Also in spring and summer, feed them once a month with half-strength balanced fertilizer. The most important thing is to protect these plants from direct sunlight and to keep the air around them moist by regular misting: placing the container on a tray filled with moist clay granules will help to raise the humidity levels.

If you're keen to propagate more plants, simply keep your plant in a container that's slightly too large for it, and the creeping stems will root in the compost. These can easily be snipped off and grown on as new plants.

Remove the insignificant flowers to encourage the plant to put all its energy into producing its attractive foliage.

PROBLEM SOLVING It's not unusual for the mosaic plant to suddenly die. This often happens if plants have been overwatered or subjected to cold in winter. Yellowing leaves indicate that the environment needs adjusting, so take action before it's too late.

If your plants become straggly, remedy this this by pruning them, which will encourage new growth.

DISPLAY

Every home should have at least one mosaic plant. With its stunning leaves, this small plant adds colour to a mixed planting and can be matched to the theme of a room. To enjoy the plant at its best, plant in a terrarium or a bottle garden with a wide opening. Put a layer of gravel in the bottom for drainage, quarter-fill with multipurpose compost and plant with other small plants that require a humid home, such as the maidenhair fern (see p.50) and the radiator plant (see p.117).

Loving high humidity, the mosaic plant will flourish in a bottle garden.

ALSO TRY

Houseplants that have colourful leaf veins are almost always highly prized. The contrast in colour offered by this type of foliage will introduce a lavish look to any interior. Try this equally striking plant, which will also add a touch of glamour:

• **Zebra plant** (*Aphelandra squarrosa*), height 60cm (24in). The dark green leaves of this plant have dazzling, silvery-white leaf veins. It thrives in high humidity, so it's perfect in a bathroom.

CAPE JASMINE *GARDENIA JASMINOIDES*

If you're familiar with the scent of Cape jasmine flowers, you'll know why this is such a desirable houseplant. These summer blooms resemble pure white roses. Enjoying high humidity and filtered sun, this plant is ideal for a bathroom or kitchen, but it's often placed in hallways to greet guests.

HEIGHT 60cm (24in)
SPREAD 60cm (24in)
FLOWERS Scented, white
FOLIAGE Glossy, dark green
LIGHT Filtered sun
TEMPERATURE 16–24°C (61–75°F)
CARE Challenging
PLACE OF ORIGIN South China and Japan

CARE

Cape jasmines have the reputation of being tricky to grow, but they're well worth the effort as the flowers and glossy foliage are breathtaking. The secret to success is to place this plant in the correct environment, which is a light room where it can be protected from direct sunlight – a west-facing room is ideal. Draughts and fluctuations in temperature can prove fatal to this hugely sensitive plant.

The Cape jasmine is an acid-lover, so grow it in an lime-free (ericaceous) compost – if you use multipurpose the leaves will soon start to turn yellow. Keep the compost moist during the spring and summer months; reduce the watering slightly in the winter.

Cape jasmines dislike hard water, so for healthy plants, use room-temperature rainwater. Feed every two weeks in spring and summer with half-strength, acid-loving plant fertilizer.

Keep humidity high by placing the pots on saucers of damp, expanded clay granules. Misting is also beneficial, but don't do this when the plants are in flower as it will damage the blooms.

PROBLEM SOLVING Yellowing leaves are often the result of too much or too little water, or could also be caused by using hard tap water.

If the plant isn't producing flower buds in spring, the room may be too hot and dry at night. To induce flowers later in the year, place the plant in a room that has a night-time temperature of 18°C (64°F).

DISPLAY

When in flower, this plant needs no other companions – its scent will draw your houseguests towards it. However, if the thought of a floral houseplant display appeals, group it with other flowering plants that enjoy filtered sunlight and high humidity, such as the tail flower (see p.55) and the rose of China (see p.98).

DID YOU KNOW? THE SCENT FROM THE CAPE JASMINE FLOWER IS INCLUDED IN MANY PERFUMES, LOTIONS, AND SCENTED CANDLES.

The glorious summer blooms will fill a sunny room with perfume.

ALSO TRY

Filling your home with glorious natural fragrance is immensely appealing. There are a number of other houseplants, in addition to Cape jasmine, that are well worth growing for the perfume of their scented flowers as well as their beauty. You'd be well advised to avoid perfumed plants in bedrooms (many people find them overpowering), but consider placing this plant in other rooms or areas around your home:

• **Cherry pie** (*Heliotropium arborescens*), height 60cm (24in). This tender perennial is often also grown as an annual in the garden, but display the seductive, dark blue, flowering plant inside and it will fill a room with scent.

Shiny leaves are a magnificent backdrop for the Cape jasmine's pure white flowers.

RATTLESNAKE PLANT

GOEPPERTIA LANCIFOLIA

HEIGHT 75cm (30in)
SPREAD 45cm (18in)
FLOWERS Insignificant
FOLIAGE Wavy edged
LIGHT Filtered sun/light shade
TEMPERATURE 16–24°C (61–75°F)
CARE Fairly easy
PLACE OF ORIGIN Brazil

The foliage of this plant is exceptional and will add a modern look to an interior. The top side has snake-like markings in two tones of green and the underside is burgundy. Enjoying moderate humidity and filtered sun, this plant is at home in a west-facing room, especially a humid bathroom or kitchen.

CARE

For success, grow your rattlesnake plant in a bright room where it can be protected from direct sunlight – too much sun will cause the striking foliage markings to fade.

Plant in multipurpose compost and keep consistently moist all year round. This plant thrives when temperatures and watering regimes don't fluctuate throughout the year. To keep the humidity high around the plant, place the pot on a tray of damp, expanded clay granules and mist it regularly.

Feed with a balanced liquid fertilizer every month in the growing season to encourage flowers.

Prune out any damaged leaves right at the base or neatly trim off brown tips. Rattlesnake plants will enjoy renewed vigour if they're repotted in early summer; this is also the best time to divide the rattlesnake plant to create new plants.

PROBLEM SOLVING Red spider mites (see p.47) are attracted to this plant, so be sure to regularly check the undersides of leaves. The edges of the foliage can sometimes start to wither. The likely cause of this is too much calcium in the tap water. To remedy, try watering with room-temperature rainwater.

DISPLAY

With its eye-catching, distinctively patterned long leaves, the rattlesnake plant is a spectacular addition to any interior and will introduce a contemporary, tropical feel. A good specimen offers a well-balanced shape, which makes it suitable for displaying on its own in the centre of a table. For a truly tropical display, group it with plants that have contrasting shape and texture but are similar colour, such as the velvet plant (see p.95) and the radiator plant (see p.117).

ALSO TRY

Foliage plants don't get much more interesting than the rattlesnake plant. Offering both patterned foliage and a wavy edge, it's consistently popular. If you've fallen for its charms, consider trying other goeppertias, such as:

• **Peacock plant** (*Goeppertia makoyana*), height 60cm (24in). This plant has silver leaves with dark green markings and burgundy markings beneath.

The peacock plant will thrive in filtered sun and high humidity.

The front and back of rattlesnake plants' leaves are completely different.

SCARLET STAR *GUZMANIA LINGULATA*

This eye-catching bromeliad has dark green, shiny foliage and tiny, white flowers presented at the top of a stem made up of red bracts. Scarlet star is just the answer if you're keen to introduce a burst of powerful colour into a room. The plant flourishes in high humidity and filtered sunlight.

HEIGHT 45cm (18in)
SPREAD 45cm (18in)
FLOWERS Small, white
FOLIAGE Glossy, green
LIGHT Filtered sun
TEMPERATURE 18–27°C (64–81°F)
CARE Challenging
PLACE OF ORIGIN Central America

Scarlet star, like all the bromeliads, will die soon after flowering.

CARE

In its native rainforest home, scarlet star grows on the branches of larger plants. So, with no need to search for moisture, its roots are small and it happily grows in a small container. However, humidity is essential – the neat, low-growing plant is therefore ideally placed in a bathroom.

Pot into multipurpose or orchid compost. Water directly into the centre of the urn with rainwater. Tip out and replace the water in the urn every week and keep the surface of the compost just moist. Feed the plant with a balanced fertilizer once a month during the spring and summer months.

As with all bromeliads, once the long-lasting flower spikes of the scarlet star have faded the plant will start to die. However, there's no need for alarm as you'll find new plants forming at the base of the parent plant.

If scarlet star is seen from above, its name makes perfect sense.

PROBLEM SOLVING If the leaves of the plant start to turn yellow it's likely to be due to a lack of light, so move your scarlet star to a sunnier spot.

If you forget to change the water in the urn, the plant may suffer from rot at its heart or it may start to smell of stale water. Reduce watering for a while to try to resolve these issues.

DISPLAY

Scarlet star is popular with interior designers thanks to its impressive, bright red bracts and neat, architectural shape. It makes a striking addition to a mixed bowl of houseplants. As guzmanias mustn't be overwatered, if planting in a group display, sink its pot into the compost of a larger bowl so you can water it differently from the other plants. Grow with high-humidity lovers such as maidenhair fern (see p.50) and Chinese evergreen (see p.52).

ALSO TRY

Bromeliads with colourful bracts are a wonderful way of introducing long-lasting and powerful colour to a room. Also consider this eye-catching plant for its bracts:

- **Flaming sword** (*Vriesea splendens*), height 60cm (24in). The flower spike of this popular bromeliad is accentuated with magnificent, sword-shaped scarlet bracts. The plant also produces small, tubular, greenish-yellow flowers, but it's undoubtedly the bracts that steal the show.

Flaming sword's stunning bracts are offset by arching, striped leaves.

VELVET PLANT
GYNURA AURANTIACA

The leaves of this plant are covered in fine, purple hairs, making them velvety to the touch. With a metallic green topside and dark purple underside and stems, the plant has a rich, luxurious appearance. It thrives in moderate humidity and filtered sunlight and is not a hard plant to place.

HEIGHT 20cm (8in)
SPREAD 20cm (8in)
FLOWERS Orange/yellow
FOLIAGE Green and purple, velvety
LIGHT Filtered sun
TEMPERATURE 15–24°C (59–75°F)
CARE Fairly easy
PLACE OF ORIGIN Southeast Asia

The leaves of this plant have a wonderful velvety texture.

CARE

Filtered or bright, indirect light is required to keep the colour of the velvet plant strong and vigorous.

Grow this plant in a generous container filled with an equal mix of multipurpose and loam-based compost. Keep the plant moist and also apply half-strength, balanced feed in the spring and summer. Reduce watering slightly during the winter and also stop feeding.

The foliage will look good only if it's always kept dry. If the leaves turn brown (indicating low humidity), don't resort to misting the velvet plant. Instead, place it on a dish filled with soaked clay granules.

Regularly pinch out (pinch off with your fingers or scissors) the growing tips to keep it bushy. If flowers start to form, cut them off as they have an unpleasant smell.

PROBLEM SOLVING This velvet plant is seldom affected by pests and diseases but it's common for the foliage to lose its appealing colour. This is due to light levels being too low – remedy this by moving the plant to a sunnier spot. Overwatering can cause rot.

DISPLAY

This plant is quite capable of attracting attention all on its own, especially in a purple or pink container. With its fabulous, velvety leaves, it's a terrific conversation piece, making it an ideal plant for a coffee table.

To create a mixed display in one pot, pair it with neatly shaped, smaller plants that enjoy the same conditions, such as the mosaic plant (see *p.91*) or the radiator plant (see *p.117*).

The flowers look fantastic, but their smell is unpleasant.

ALSO TRY

A good collection of houseplants should include a mixture of colours, textures, and shapes. If the idea of tactile plants appeals to you, then why not consider these two plants:

• **Moonstones** (*Pachyphytum oviferum*), height 10cm (4in). This plant has plump, rounded succulent leaves that are as soft as a smooth pebble to the touch.

• **Spear sansevieria** (*Sansevieria cylindrica*), height 75cm (30in). The smooth, cylindrical leaves of this succulent plant have a wonderful structure.

Spear sansevieria occasionally produces fragrant flowers.

ZEBRA CACTUS

HAWORTHIA FASCIATA

This spiky, stripy succulent has a modern look and works well in a sunny kitchen, where it fits right in alongside stainless steel. The fact it needs very little attention adds to its appeal, making it a popular choice for a student flat or for placing on the office desk.

HEIGHT 20cm (8in)
SPREAD 15cm (6in)
FLOWERS Tubular, white
FOLIAGE Succulent, striped
LIGHT Sun/filtered sun
TEMPERATURE 12–26°C (54–79°F)
CARE Easy
PLACE OF ORIGIN Eastern Cape of South Africa

CARE

The zebra cactus thrives on neglect, so it's a terrific choice for anyone who doesn't want to be tied down by responsibilities towards their houseplants. The plant is happy to live on a sunny windowsill, which means that you can find room for it in even the smallest of homes.

Plant in a container of cactus compost that will offer just a little bit more room for root growth. Watering requirements for this plant are minimal: allow the compost to dry out before rewetting. Apply a cactus plant food in the spring every two weeks until autumn arrives.

Very mature zebra cacti produce tubular, white flowers in the summer months. Repot every other year during the spring and take this opportunity to remove offsets that can be grown on as new plants.

PROBLEM SOLVING Giving the zebra cactus too much attention is often the cause of its demise. Overwatering the plant, especially in winter, when temperatures and light levels have dropped very low, can be fatal. If you're known to be too generous with your watering, a container that has good drainage holes is absolutely vital, otherwise the leaves will soon start to lose their healthy glow.

DISPLAY

With its beautifully marked, slender stems and manageable size, this plant is great for cutting-edge interiors. Its spiky appearance also brings a sense of fun. If you have a sunny room with low humidity, display your zebra cactus with other sun-loving succulents such as echeveria (see p.83) and bunny ears cactus (see p.115) in a dry terrarium. Choose a glass pot or specially designed terrarium with an open side (a bottle will become too humid). Place a layer of gravel in the bottom and add cactus compost before planting. Alternatively, plant in a container on its own – grey pots can look great with this plant.

It's almost impossible to kill the zebra cactus, unless you overwater it.

Grow your zebra cactus with other small succulents in a dry terrarium.

ALSO TRY

When space is an issue, a windowsill is a great place for sun-loving plants. To avoid blocking out the light, small plants are the answer. Try combining these succulents with zebra cactus:

- **Beautiful graptopetalum** (*Graptopetalum bellum*), height 15cm (6in). This plant has tight, neat rosettes of grey leaves.
- **Donkey's tail** (*Sedum morganianum*), height 10cm (4in). This plant has rope-like stems with tiny, round leaves.

IVY
HEDERA HELIX

There are many forms of *Hedera helix* and their vigour, leaf colour, and size vary greatly. However, all of them are easy to grow and can cope with cool, unheated rooms. Don't dismiss ivies as houseplants – they deserve to be celebrated and can grow where others will not.

HEIGHT 1.5m (5ft)
SPREAD 90cm (36in)
FLOWERS Insignificant
FOLIAGE Variegated or plain green
LIGHT Filtered sun/shade
TEMPERATURE 5–20°C (41–68°F)
CARE Easy
PLACE OF ORIGIN Most of Europe

Choose plain green or variegated ivy to trail or climb in your home.

CARE

Most modern homes have central heating and are well insulated, but if you're looking for a houseplant that copes in cold rooms, this plant is ideal: a cool but frost-free location is perfect for ivy. It not only puts up with changes in temperature but also tolerates draughts.

Plant in multipurpose compost. In the growing season, keep this just moist, but reduce watering slightly in winter. Never let ivy dry out. Your plants will manage without regular, additional feed but providing them with application of a balanced fertilizer every month in spring and summer will certainly improve their performance.

Keep plants looking compact and neat by pruning to shape as and when required and repot every other year to improve their health and vigour.

PROBLEM SOLVING Variegated plants have a tendency to revert to green. This is easily solved by cutting out the stem to prevent green leaves from taking over, then moving the plant to a position that receives more light.

If the leaf tips of your ivy turn brown it's most likely that the air is too dry, so treat the plant to a regular misting.

DISPLAY

Smaller forms of ivy are perfect for the edge of a generous container, while larger types are ideal for covering trellis or trailing from a hanging basket. If the plant is allowed to climb, it will attach to a wall by itself – in an interior, use a trellis to protect your walls.

You can have tremendous fun training ivy over supports in your home. Buy shaped training supports that are designed to push into your container and then wind the growth into it. Other climbing plants such as the rose grape (*see p.108*) and stephanotis (*see p.134*) can also be trained in this way and displayed with ivy.

ALSO TRY

Growing climbing/trailing plants in hanging baskets is increasingly popular. Ivy is often displayed in this way, but if you're looking for other trailing, basket plants, try:

- **String of turtles** (*Peperomia prostrata*), trail 70cm (28in). This plant has flat, round leaves decorated with a marble pattern.
- **Teddy bear vine** (*Cyanotis beddomei*), trail 60cm (24in). This plant is named for its soft and hairy leaves, which have an attractive, maroon underside.

String of turtles will thrive in a sunny room with low humidity.

ROSE OF CHINA

HIBISCUS ROSA-SINENSIS

The much-celebrated rose of China is grown for its exquisite (albeit fleeting) blooms. These are large, tropical, and come in a range of colours, including red, orange, and white. A spectacular dark red throat sits at the centre of most flowers, adding to the drama and impact.

HEIGHT 1m (3ft 3in)
SPREAD 2m (6ft 6in)
FLOWERS Large, trumpet-shaped
FOLIAGE Glossy, green
LIGHT Sun/filtered sun
TEMPERATURE 10–26°C (50–79°F)
CARE Challenging
PLACE OF ORIGIN China

Rose of China blooms are among the most stunning of all flowering houseplants.

CARE

Finding the right position in the home for this tropical beauty can be tricky, but if you get it right you'll be rewarded with magnificent blooms during the summer. Be aware that these may only last a day in a warm room. The rose of China needs bright, humid conditions, out of direct sunlight and with good ventilation but away from draughts in the winter. A conservatory is often the perfect place for the plant as the doors can be flung wide open when the weather is warm.

Grow in multipurpose compost and keep moist in spring and summer, but reduce watering in winter. Applying a balanced fertilizer every two weeks in spring and summer will increase the chances of flowers. To keep the plant neat and manageable, prune back by a third in spring.

PROBLEM SOLVING Rose of China is primarily grown for its flowers, so the disappointment when buds drop before opening is huge. This tends to happen if they've become too dry over summer,

Each flower has five petals and may only last a single day.

if you've moved the plant to a new location, or if the temperature has changed dramatically.

DISPLAY

Plants are often bought as fairly small specimens for display on tabletops. It's rare they'll reach 1m (3ft 3in) in height in the home as most indoor gardeners prune them to a more manageable size. When in flower, this plant is the perfect solo performer. While waiting for flowers to form, couple the hibiscus with other flowering plants – these will thrive under the same conditions and, with luck, flower at different times. Try the rose grape (see *p.108*) and stephanotis (see *p.134*).

ALSO TRY

If the thought of a flower-filled house appeals and you have a room with filtered sun and moderate humidity, then why not add more tropical-looking blooms to your collection with this plant:
- **Persian violet** (*Exacum affine*), height 20cm (8in). If space is tight, this is the ideal flowering plant. It has violet-blue flowers with a yellow centre and is a biennial, so expect flowers in the second year.

AMARYLLIS *HIPPEASTRUM*

This striking plant is grown for its giant, trumpet-shaped flowers that appear year after year. A wide range of colours is available, including red, orange, and pink, any of which will create a splash of colour in your home. The long, wonderfully glossy leaves have an elegant, arching habit.

HEIGHT 60cm (24in)
SPREAD 30cm (12in)
FLOWERS Large, trumpet-shaped
FOLIAGE Insignificant
LIGHT Filtered sun
TEMPERATURE 13–21°C (55–70°F)
CARE Fairly easy
PLACE OF ORIGIN Brazil

CARE

Amaryllis plants should be placed in a bright, warm room but out of direct sunlight. As soon as flowers start to appear, move to a cooler room so that the blooms will last for longer.

If you buy as a bulb between October and January, plant it into multipurpose compost in a pot that's only slightly larger than the bulb (they flower best in small pots). One third of the bulb should sit above the compost surface. Water sparingly until growth appears and feed every 10 days with a tomato fertilizer to encourage flowers.

Buy and plant amaryllis bulbs in autumn or early winter.

After flowering, deadhead and cut off any faded foliage, reduce watering, and move the plant to a cool but bright, frost-free garage or shed for two months before moving back inside. Increase the watering to start them back into growth. Don't be tempted to repot your amaryllis too often – once every three years is perfect as they dislike any disturbance.

Older bulbs will produce offsets that can be carefully removed and potted on to create new plants.

PROBLEM SOLVING The impressive flowers of this plant tend to lean towards the light; avoid this by turning the plants daily. Failure to flower could be due to not enough light or underwatering during the previous summer. Amaryllis are sometimes susceptible to bulb pests such as narcissus bulb fly.

If the impressive flowering stems start to flop, support them with a cane.

DISPLAY

As amaryllis require turning, they should be planted in individual pots – the more decorative, the better. If in flower at Christmas they make the perfect festive table centre. For best effect, display in a group with other amaryllis hybrids of different colours.

ALSO TRY

There are many other bulbs that can bring impressive flowers into your home. They're often bought to celebrate a special event such as Christmas or Easter. A plant that's frequently grown at Easter is:

• **Grape hyacinth** (*Muscari*), height 20cm (8in). This plant's spires of blue flowers look great planted in a group display in a pot. They're often grown outside but add spring colour to a cool room.

Plant grape hyacinth bulbs in autumn for a spring display.

KENTIA PALM *HOWEA FORSTERIANA*

With its eye-catching, elegant foliage, this plant will add height and tropical glamour to a room. The kentia palm is a popular choice as it forms a very upright plant, unlike many other palms, which have a more arching form – its habit makes it easier to fit into a small room.

HEIGHT 3m (10ft)
SPREAD 2m (6ft 6in)
FLOWERS Rarely blooms indoors
FOLIAGE Dark green
LIGHT Light shade
TEMPERATURE 13–24°C (55–75°F)
CARE Fairly easy
PLACE OF ORIGIN Australia

Kentia palms have great presence and impact when displayed in groups of three.

CARE

The kentia palm is a problem-solver: if you have a room with a bare corner in moderate humidity that receives little sunlight, this palm will thrive there.

Plant in a 50:50 mix of multipurpose compost and loam-based compost in a heavy, generous container (large plants can become top-heavy). Only water the plant when the top of the compost is slightly dry and avoid overwatering it in the winter. Feed with a balanced liquid feed every fortnight during spring and summer.

The kentia palm is popular for its tolerance of neglect. To keep this easygoing plant happy, simply mist it regularly, keep away from draughts, and prune out only damaged or yellowing leaves. This is a slow-growing plant, so you won't need to repot it more than every three years.

PROBLEM SOLVING Brown tips on the leaves are often caused by draught or dry air. Simply snip them off, move the plant, and mist the foliage more regularly. The kentia palm is prone to attack from red spider mites and mealybugs (see pp.46–47). If the red spider mites are only affecting one leaf, remove the leaf to curb the spread.

DISPLAY

Put some thought into where you place your palm. These are structural feature plants and, in a large home, a pair of them will look stunning framing a doorway. Small specimens are popular tabletop plants and larger specimens will create a focal point. Invest in one large specimen, and a room can soon be transformed into a tropical paradise. If growing in a large container, consider planting the base with trailing or smaller foliage plants that also tolerate shade, such as devil's ivy (see p.84) and Boston fern (see p.112).

ALSO TRY

Palms add height, glamour, and a tropical touch. If you have a room with filtered sunlight, the kentia palm may not be suitable. But there are other striking palms that prefer higher light levels, including:

- **Areca palm** (*Dypsis lutescens*), height 2m (6ft 6in). This plant has a more of an arching habit than the kentia palm but a very similar look.
- **Fishtail palm** (*Caryota mitis*), height 2.5m (8ft). This plant's upright habit and unusual fronds are made up of striking, fishtail-shaped foliage.

The areca palm looks best when its arching stems are given space.

WAXFLOWER *HOYA*

This plant is grown for its trailing habit and heads of scented, white flowers that fill a small room with perfume on summer evenings. Waxflower is perfect for a space that's well lit and has moderate humidity, such as a sunny kitchen, where it can trail from a hanging basket or tumble from a windowsill pot.

HEIGHT 4m (13ft)
SPREAD 4m (13ft)
FLOWERS Waxy, white, scented
FOLIAGE Dark green
LIGHT Filtered sun
TEMPERATURE 16–24°C (61–75°F)
CARE Fairly easy
PLACE OF ORIGIN China and Indonesia
WARNING! Sap is toxic; gloves required when handling

CARE

Plant waxflower in an equal mix of orchid compost, multipurpose compost, and perlite as it will only thrive in a well-aerated growing medium. Keep the compost moist in spring and summer, but reduce watering in winter. Feed the plant with a half-strength, balanced fertilizer between spring and autumn and increase the humidity around it by placing the container on a tray filled with damp, expanded clay granules.

After the flowers have faded, don't be too quick to deadhead them as they may rebloom from the base of the old flowers. Be cautious when

Waxflower will climb a support, but is most often grown as a trailing plant.

Expect the scented flowers to appear during the summer months.

it comes to pruning this plant, too – a light pruning will often produce better results than hard pruning.

PROBLEM SOLVING If the leaves start to blacken, this is likely to be the result of trying to grow the plant in the wrong compost. Repotting is done in the same way as planting (*see left*).

It can be tempting to try to urge flowers to form by overfeeding, but this will cause the plant to die back. Encourage flowering by moving it to a brighter spot, but avoid direct sunlight.

DISPLAY

Scented flowers appear on the waxflower from late spring to early autumn, but even without them, this is a useful climbing or trailing plant. If your aim is to cover a trellis or wall,

choose *Hoya carnosa*, while hanging baskets are the perfect home for trailing waxflowers. A series of them hung artistically together can give the effect of a green wall or hanging garden. Other trailing plants that enjoy the same conditions as the waxflower, and will partner well with it, include hearts on a string (*see p.63*) and the silver inch plant (*see p.139*).

ALSO TRY

Displaying houseplants in hanging baskets is popular for bathrooms. Like the waxflower, any plants you choose must be able to cope with moderate levels of humidity. If planning a display to enjoy while relaxing in the bath, try this plant:
• **Creeping Charlie** (*Pilea nummulariifolia*), height 30cm (12in). This creeping plant has small, green leaves that are held on fleshy, pink stems.

Creeping Charlie is easy to grow and puts on growth quickly.

POLKA DOT PLANT

HYPOESTES PHYLLOSTACHYA

This joyful little plant offers vibrant foliage and a neat, compact shape. Its green leaves are speckled with pink, red, or cream – combine different coloured plants to add real sparkle to your home. The polka dot plant likes indirect light, but too little light will cause its leaves to fade.

HEIGHT 25cm (10in)
SPREAD 25cm (10in)
FLOWERS Insignificant
FOLIAGE Colourful, speckled
LIGHT Filtered sun
TEMPERATURE 18–27°C (64–81°F)
CARE Fairly easy
PLACE OF ORIGIN South Africa

If you're short on space, you'll always find room for this neat little plant.

CARE

The polka dot plant likes filtered sun and moderate humidity, so it will thrive in a kitchen or bathroom, or in any room if grown in a humid bottle garden or terrarium. Plant in multipurpose compost with a little perlite mixed in for drainage. Let this dry out slightly before rewetting. Reduce watering in winter; never overwater if growing in a pot without drainage holes, such as a bottle garden. Feed the plant with a balanced fertilizer every two weeks throughout the spring and summer.

Mist plants regularly and occasionally pinch out (take off with your fingers or scissors) the growing tips to keep the plant compact. Some gardeners even remove the small purple flowers as they can spoil the shape of the plant.

PROBLEM SOLVING The colours in the foliage can fade if the plant isn't in a bright enough place – if this happens, simply move it to a brighter spot. Scale insects, whitefly, and mealybugs can also cause issues (see pp.46–47). If growing the polka dot plant in a bottle garden, act quickly, otherwise the pests will soon take over.

DISPLAY

Although small, the polka dot plant creates a big impression, thanks to its vibrant foliage – the colours pop and will add sparkle to a sunny room. Place it so that it can be easily viewed from above – you won't get the full impact of the plant if it's on the top shelf. As this plant remains fairly small and neat you can have great fun experimenting with your choice of pot. For example, consider planting it in a large, decorative mug. Alternatively, it's equally striking in a pot on the windowsill or as the lead plant in a bottle garden – children love planting these and in most garden centres you can buy small, pocket-money plants to create them. Other great candidates for this kind of display include the maidenhair fern (see p.50) and the mosaic plant (see p.91).

(see pp.46–47) ... (see p.50) ... (see p.91)

ALSO TRY

Decorative foliage is popular, and it can be great fun matching foliage to other features in a room, such as cushions, carpets, and curtains. If you've fallen for the polka dot plant and want more plants with colourfully variegated foliage, try:

- **Polka dot begonia** (*Begonia maculata*), height 40cm (16in). This plant has green and white spotted leaves with cascading cream flowers. It enjoys light shade.

The polka dot begonia likes high humidity, but never mist its leaves.

JASMINE *JASMINUM POLYANTHUM*

This stunning climber has small, dark green leaves and panicles of highly scented, white flowers that start as pink buds. Jasmine's exceptional blooms will appear in midwinter and last a month or more, providing that feel-good factor when the days are short, dark, and cold.

HEIGHT 3m (10ft)
SPREAD 3m (10ft)
FLOWERS White, scented
FOLIAGE Dark green
LIGHT Filtered sun
TEMPERATURE 10–24°C (50–75°F)
CARE Fairly easy
PLACE OF ORIGIN Southwest China

CARE

Jasmine enjoys low humidity and filtered sunlight, making it ideal for a cool room or conservatory but inappropriate for rooms with central heating, which is too dry. If growing in a room other than a conservatory, give it plenty of space and something to climb up; if growing in a conservatory, train it along galvanized wires attached to the house wall.

Plant jasmine in a generous container, especially if growing it up a wall as repotting will be tricky. It's far easier to remove tired compost from the top of the pot in spring and replace it with fresh compost. A loam-based compost with a few handfuls of perlite is ideal. Keep the compost moist in the spring and summer and reduce a little in winter, but if you see flower buds forming, increase watering slightly again. Applying balanced liquid fertilizer every two weeks between spring and autumn will encourage the plant's much-admired blooms. Jasmine is a vigorous plant – keep it to a manageable size by cutting back after flowering.

PROBLEM SOLVING Jasmine is rarely affected by pests and diseases. The foliage will start to look tatty and turn yellow if it's underwatered, not given enough light, or if the plant gets too hot. Dry air from central heating will also affect the health of the foliage. Cool nights, such as those offered by an unheated conservatory, are perfect.

DISPLAY

If growing this plant in a home that you don't intend to stay in for long, train it up a wire frame that slots into the pot rather than training up a wall. A wall can be covered in foliage fairly quickly, especially if you plant your jasmine with other plants that enjoy low humidity and filtered sunlight, such as grape ivy (see p.66) and devil's ivy (see p.84).

Jasmine buds are pink, but the results are pure white blooms.

A happy jasmine will be covered with scented blooms in winter.

ALSO TRY

Many plants thrive in a cool room or conservatory. If you've had success with jasmine and are looking for other exciting plants to accompany it in your garden room, then consider these two:
- **Chilean bellflower** (*Lapageria rosea*), height 3m (10ft). This climber with pink, bell-shaped flowers will suit cool, shady areas.
- **Spider lily** (*Hymenocallis speciosa*), height 60cm (24in). The white, star-shaped flowers of this bulbous perennial are magnificently perfumed.

SHRIMP PLANT *JUSTICA BRANDEGEEANA*

HEIGHT	90cm (36in)
SPREAD	90cm (36in)
FLOWERS	Small, white
FOLIAGE	Dark green
LIGHT	Filtered sun
TEMPERATURE	15–25°C (59–77°F)
CARE	Easy
PLACE OF ORIGIN	Mexico

This plant is grown for its unusual flowers, which are present almost all year. The white blooms are enclosed in pink and yellow bracts – when in flower, it's clear why it's called the shrimp plant. It's ideal for a sunny sitting room with low humidity, where you can sit back and admire its charms.

CARE

The shrimp plant isn't fussy, but it does require warm days and cooler nights for success. If you turn your heating right down at night, this could be an ideal plant for you. It has the form of an upright shrub and, over time, will make an impressive specimen.

Plant in a loam-based compost. This should remain moist throughout the spring and summer months. In winter, reduce watering and give the compost a chance to dry before you water again. To help with vigour and encourage a good supply of flowers, feed every two weeks in spring and summer with a balanced liquid fertilizer.

To prevent the colours of the bracts from fading, protect the shrimp plant from direct sunlight. Give it a hard prune during the spring, cutting back

The shrimp plant will thrive in a sunny room with low humidity.

Small white flowers can be seen at the end of the salmon-pink bracts.

to a third of its size to encourage healthy new shoots and create a compact, neat plant. Use some of this plant material to take cuttings – a homegrown young shrimp plant makes a great gift.

PROBLEM SOLVING Red spider mites (see p.47) are often a problem with the shrimp plant. You'll notice the mottling on the leaves before the tiny pests. Remove any affected leaves or, if the case isn't severe, try misting the plant with water. However, if the spider mites have really taken hold, you may need to turn to a pesticide.

DISPLAY

The shrimp plant enjoys colder nights, so it's often grown in a frost-free conservatory with other plants that can cope with a drop in temperature after dark, such as jasmine (see p.103) and scented-leaved pelargoniums (see p.116). Both these companions will add magnificent fragrance to a group display.

ALSO TRY

Long-lasting flowers are very appealing, especially if you don't have the room for a large collection of plants. If the thought of a persistent floral display appeals, then this is also a good option:

• **Yellow sage** (*Lantana camara*), height 1m (3ft 3in). This plant blooms from spring to autumn in a range of colours, including red, pink, yellow, orange, and cream.

Put yellow sage in a sunny spot: it loves a Mediterranean climate.

FLAMING KATY

KALANCHOE BLOSSFELDIANA

If you're looking for a plant that brings cheer to a room and requires very little in the way of care, this is it. Flaming Katy is neat, easy to place, and tremendous value as the flowers last up to three months. It is, however, a short-lived plant and rarely reflowers, so expect to replace it each year.

HEIGHT 45cm (18in)
SPREAD 30cm (12in)
FLOWERS Red, pink, orange, or white
FOLIAGE Dark green, succulent
LIGHT Sun/filtered sun
TEMPERATURE 12–26°C (54–79°F)
CARE Easy
PLACE OF ORIGIN Madagascar

CARE

Flaming Katy is an ideal plant for a sunny windowsill in a room with low humidity, such as a centrally heated sitting room, where the air is dry.

When watering the plant from above, don't allow the water to touch the leaves as this can lead to rot. Compost should only be slightly moist and almost dry in winter. You can apply a half-strength balanced liquid feed in spring and summer, but it isn't essential. There's no need to repot this plant if you're only keeping it for one year.

However, if you fancy the challenge of keeping a flaming Katy for another year, cut off the faded flower stems and move the plant to a shadier spot to rest. Repot it in cactus compost after flowering. This plant can be easily propagated by taking leaf cuttings or stem cuttings during the spring or summer months (see pp.38–39).

PROBLEM SOLVING Kalanchoes are rarely affected by pests and diseases but it's not unheard of for them to attract mealybugs (see p.46). Check for these pests under leaves or between leaf axils. They're easy to spot as they excrete a white, fluffy substance that will stand out clearly on the foliage.

DISPLAY

With its vivid winter flowers, flaming Katy is a hugely popular houseplant for introducing festive colour into homes during both Christmas and Chinese New Year celebrations.

The plant's shiny, dark green leaves provide a wonderful backdrop to the colourful red, pink, orange, or white stems of flowers. You can create a vibrant display in your entrance hall or sitting room by placing flaming Katy with other seasonal flowering plants, such as cyclamen (see p.74) and the traditional Christmas houseplant poinsettia (see p.85).

Plants are often discarded once the flowers have faded.

ALSO TRY

Succulents are perfect for anyone who may be too busy to tend to their houseplants regularly, or is away from their plants a lot. Try these other flowering succulents if you're worrying about not being around to care for your plants:

- **Flower dust plant** (*Kalanchoe pumila*), height 45cm (18in). This plant has fine silver foliage and very pretty, pale pink flowers.
- **Rat's tail cactus** (*Disocactus flagelliformis*), height 60cm (24in). Perfect for a hanging basket, the rat's tail cactus has flat leaves with stunning red blooms.

The attractive, double-flowered form of flaming Katy is shown here.

LIVING STONE *LITHOPS*

These sun-loving succulents, which resemble pebbles, make great pocket-money plants for kids – they require little in the way of care, take up hardly any space, and look very different from other houseplants. With so many species available, living stones can soon become addictive collectables.

HEIGHT 5cm (2in)
SPREAD 5cm (2in)
FLOWERS Daisy-like
FOLIAGE Succulent
LIGHT Sun
TEMPERATURE 18–26°C (64–79°F)
CARE Easy
PLACE OF ORIGIN South Africa

A south-facing windowsill is the most suitable home for living stones.

CARE

Sun is important to these pebble-like plants, so a south- or west-facing windowsill is ideal. Find a hot spot that's draught free, and you can almost ignore them. As living stones prefer low humidity, avoid placing them in a kitchen or bathroom – their natural home is desert or semi-desert, so a moist environment won't suit them at all. The living stone's succulent body is designed to hold water, which means it has its own storage tank to rely on.

These plants rarely need repotting, but if you do need to do this, choose a cactus compost and add a generous sprinkling of horticultural sand to improve the drainage. Growing in a clay pot is a good idea as it will lose

moisture quicker than a plastic pot, so it's the best choice if you tend to overwater. Water very lightly in spring and summer; reduce to almost nothing in winter. Feeding isn't necessary.

The daisy-like flowers – most often in yellow or white – appear in late autumn. A plant needs to be a few years old to flower, so be patient.

PROBLEM SOLVING Overwatering will cause your lithops to rot, so water with caution. After the flower has faded, leaves start to wither away, which can cause concern. However, this is quite natural and nothing to worry about as two new fused leaves will soon appear between the two old leaves.

The daisy-shaped autumn flowers are best viewed from above.

DID YOU KNOW? LITHOPS HAVE EVOLVED TO RESEMBLE STONES SO THAT ANIMALS WON'T EAT THEM IN THEIR NATIVE HABITAT.

DISPLAY

A collection of living stones on your windowsill will look spectacular and will almost undoubtedly generate interest. As they're such small plants it would be easy to swamp them with larger plants, so be sure to select their companions carefully. It's always a good idea to combine them with other small succulents such as echeveria (see p.83) and bunny ears cactus (see p.115). All will enjoy a sunny spot.

ALSO TRY

Growing small plants that are unusual and have an interesting story behind them is a great way of getting kids into gardening. If you love the idea of your sunny windowsills covered in interesting houseplants, also consider:

• **Moonstones** (*Pachyphytum oviferum*), height 10cm (4in). This succulent plant looks just like a handful of smooth pebbles.
• **Old man cactus** (*Cephalocereus senilis*), height 30cm (12in). The fine, white hairs on this neat, column-shaped cactus resemble an elderly man's beard.

PRAYER PLANT

MARANTA LEUCONEURA VAR. *KERCHOVEANA*

This plant is popular for its beautifully patterned foliage: the light green leaves, marked with darker green blotches, resemble a feather. The prayer plant is grown as a short climber or, more often, a trailing plant. It's ideal for a room with moderate humidity and filtered light.

HEIGHT 60cm (24in)
SPREAD 60cm (24in)
FLOWERS Insignificant
FOLIAGE Feather-patterned
LIGHT Filtered sun/light shade
TEMPERATURE 15–24°C (59–75°F)
CARE Easy
PLACE OF ORIGIN South America

CARE

The prayer plant copes in light shade and is easy to place, as long as it doesn't stand in a draught. Plant in a multipurpose compost and keep moist (but not wet) in spring and summer; reduce the watering slightly in winter. Feed in spring and summer with a half-strength balanced fertilizer and stand the pot on a saucer of moist, expanded clay granules. The plant likes a fairly humid environment, so misting the leaves in a centrally heated room will also improve growth. The slow-growing plant spreads low and wide in its native tropical forest habitat. It can be left to trail slightly, or may be kept neat and compact by snipping off stems at any time of the year.

When repotting the plant every other spring, take the opportunity to divide the plants to make more.

PROBLEM SOLVING Leaf tips turning brown could be due either to red spider mites (see p.47) or to the air around the plant being too dry. If you have this issue, move the plant to a more humid room, such as a bathroom, or mist it more regularly. Carefully cut off the brown tips, leaving the foliage looking as natural in shape as possible.

DISPLAY

Display in a sitting room or bedroom and you can enjoy watching the foliage roll up during the night. The prayer plant is ideal for planting in a mixed display under larger specimens that will cast slight shade over the foliage. It's most commonly planted beneath the weeping fig (see p.87) or the Swiss cheese plant (see p.109) as these plants also enjoy moderate humidity and filtered sunlight. Both these larger specimens have dark green leaves, picking up on the darker shades in the prayer plant's beautifully patterned foliage.

Leaves fold up in the evening, which explains the plant's common name.

The herringbone plant has showy, sumptuous pink markings.

ALSO TRY

Other marantas are available and are just as easy to grow, including this striking form, which offers impressive decorative foliage:

• **Herringbone plant** (*Maranta leuconeura* var. *leuconeura* 'Fascinator'), height 30cm (12in). The herringbone plant's distinctive red veins create a feather-like pattern on the green leaves. The underside of the of the foliage is an attractive pink. As with the prayer plant, it offers a slight trial if grown in a hanging basket.

ROSE GRAPE *MEDINILLA MAGNIFICA*

When in bloom, this plant is a real showstopper. The pink flowers, held on long, arching stems, are surrounded by large, pink bracts that triple the impact. To get the most out of your rose grape, place it on a pedestal, plant in a generous hanging basket, or grow in a tall container.

HEIGHT 1m (3ft 3in)
SPREAD 1m (3ft 3in)
FLOWERS Pink
FOLIAGE Dark green
LIGHT Filtered sun
TEMPERATURE 17–25°C (63–77°F)
CARE Challenging
PLACE OF ORIGIN The Philippines

Displayed on a pedestal, the rose grape's blooms can be enjoyed to the full.

CARE

In its native home, this striking plant grows in the trees and enjoys a very humid environment and filtered sun. As a houseplant, rose grape likes high or moderate humidity and is equally happy in a bathroom or hallway.

Plant in an orchid compost and water moderately in spring and summer, reducing water slightly in the winter. Apply a half-strength, high-potash liquid feed every two weeks in summer to encourage flowers to form. As this plant enjoys high humidity, mist the foliage regularly, especially if you're trying to grow it in a room with dry air. Alternatively, keep the pot on a tray of damp, expanded clay granules.

Expect the long-lasting and highly prized flowers to appear during the spring. Once the blooms have faded, cut off the flower stems and, if you're lucky, you might get a second set of blooms. It's possible to take tip cuttings to grow on, but the success rate is low, so it would be preferable to buy another plant.

PROBLEM SOLVING Red spider mites (*see p.47*) are a common problem with the rose grape, especially if you're attempting to grow it in low humidity. You can try to prevent an infestation of these pests by misting the leaves regularly or moving the plant to a more humid room. The leaves will become mottled and may drop if the plant is left untreated.

DISPLAY

To experience rose grape's dramatic form to maximum effect, display it wherever the long arching stems, leathery foliage, and fabulous flowers can hang elegantly, as they would in their native environment. A pedestal in the middle of a room is a spectacular way to show off this decadent plant so that it can be viewed from all sides. The flowers tend to hang evenly around it, so if you position it tucked away in the corner of a room you'll be missing out on half of the show.

The rose grape is also a terrific bathroom plant and looks really impressive in the company of other plants that love high-humidity environments, such as the croton (*see p.70*) and cymbidium (*see p.75*).

(see p.47)
(see p.70)
(see p.75)

ALSO TRY

If you enjoy the appearance of the rose grape's arching habit, consider this plant, which creates a similarly impressive show:

- **Christmas cactus** (*Schlumbergera × bridgesii*), height 45cm (18in). This tropical-looking cactus has trailing stems. Buds develop in early winter and a generous amount of striking, bright pink flowers appear around Christmas time. Move the plant to a cool room when the blooms start to fade. Unlike the rose grape, Christmas cactus is an extremely easy plant to grow and maintain.

The Christmas cactus is the perfect choice for a hot, bright room.

SWISS CHEESE PLANT

MONTERA DELICIOSA

Most people are familiar with the Swiss cheese plant. Its popularity is largely due to its impressive, giant, heart-shaped, lobed or perforated leaves and its climbing habit. You'll need plenty of space for this fabulous plant, which can turn into a giant if given its ideal growing conditions.

HEIGHT 6m (20ft)
SPREAD 2.5m (8ft)
FLOWERS Unlikely to bloom indoors
FOLIAGE Large, green, lobed
LIGHT Filtered sun/light shade
TEMPERATURE 18–27°C (64–81°F)
CARE Easy
PLACE OF ORIGIN Central America
WARNING! All parts are toxic; gloves required when handling

A moss pole will offer the Swiss cheese plant the support it requires to climb.

CARE

The Swiss cheese plant likes a room with moderate humidity. If you're hoping for perforated as well as lobed leaves, place it in filtered sun rather than shade. It has aerial roots and a climbing habit, so growing it up a moss pole is the best way to offer support.

Grow in multipurpose compost and feed with a half-strength balanced fertilizer every month in spring and summer. Water when the top of the compost feels dry, reducing slightly in winter. Mist the leaves and the moss pole regularly to mimic jungle humidity and clean leaves weekly with a damp cloth to keep them healthy and shiny.

If the plant gets out of hand, simply prune off any unwanted stems and top growth during the spring. Prune just below an aerial root and this plant material can then be used as a cutting to create a new plant – just pop the cutting in water or compost.

PROBLEM SOLVING This plant is relatively pest-free. Some gardeners cut off the aerial roots – never do this as they're a crucial part of the plant and should be held against a damp moss pole or, if long enough, pushed into the compost. The leaves will be poor if these roots aren't looked after.

DISPLAY

As the Swiss cheese plant is boisterous and will quickly overshadow others, it's best suited to life on its own. If growing in a very large container, underplant it with the prayer plant (see *p.107*) or sweetheart plant (see *p.120*) – these are both shade lovers and will trail over the side of the container.

(see *p.107*) ... (see *p.120*)

ALSO TRY

Monsteras are great for those who want a large plant but are nervous about the investment – they're easy to grow and risk of failure is low. Here are two more options:

- **Mexican breadfruit plant** (*Monstera adansonii*), height 3m (10ft). The leaves of this smaller plant have intricate perforations.
- **Variegated Swiss cheese plant** (*Monstera deliciosa* 'Variegata'), height 2.5m (8ft). This plant is admired for its leaves, which are splashed with white.

This Mexican breadfruit plant has been well pruned for a tight space.

PAPER WHITE *NARCISSUS PAPYRACEUS*

HEIGHT 30cm (12in)
SPREAD 15cm (6in)
FLOWERS White, highly scented
FOLIAGE Dark green
LIGHT Filtered sun
TEMPERATURE 15–20°C (59–68°F)
CARE Fairly easy
PLACE OF ORIGIN Mediterranean

The strongest scented of all narcissus, paper whites are available as forced bulbs that will flower in time for the Christmas festivities. However, they're also often purchased, for gifts in the winter months, as short-term plants – in this case, minimal care is required.

CARE

If buying paper white plants in bud or flower in December, place them in a cool room with filtered sunlight – the cooler the room, the longer the flowers will last. Bring them into a warmer room for seasonal celebrations, then return them to their cooler spot later. Alternatively, you can force paper white bulbs yourself in September.

Winter flowers make this plant a popular gift for Christmas and New Year.

Paper whites have up to ten stunning, pure white flowers on each stem.

To do this, either plant paper whites individually or grow several of them together in one container Either way, fill the pot with multipurpose compost or bulb fibre and plant the bulbs so that the tips are just visible above the surface. Water well and place on a warm and sunny windowsill.

Whether you've forced the plant yourself or bought it in flower from a garden centre or nursery, after the flowers have faded, deadhead them but don't cut off the foliage. Remove the leaves as soon as they turn yellow. Store the plant in a frost-free place until autumn, when the bulbs can be replanted outside in a sheltered spot for a repeat display indoors. Year after year, forced or not, the bulbs will reflower in spring in the garden.

PROBLEM SOLVING If placed in a room with low light levels, the stems can become very leggy and flop as they try to search for light. To prevent this, move them to a slightly sunnier room and turn the pot regularly. You can also place a few small, green canes in the container and wrap a ribbon around them to offer the stems some support.

DISPLAY

The perfect place for a display of paper whites is in a north-facing porch, where their fragrance will fill the air and welcome your guests. They'll stay cool here and then flower for weeks. Bring them inside for special occasions and place them with other seasonal plants, such as florist cyclamen (see p.74) and poinsettia (see p.85).

ALSO TRY

Other narcissus can make great Easter dining-table decorations and will thrive for weeks in a cool room or if put outside between meals. Plant bulbs in October and keep outside until buds form. Try:
- **Double daffodil** (*Narcissus* 'Sir Winston Churchill'), height 35cm (14in). This daffodil has scented, double blooms in cream. They're quite tall, so plant bulbs in a deep container and offer a support.
- **Dwarf daffodil** (*Narcissus* 'Tête-à-tête'), height 15cm (6in). Used as a table centre, this classic daffodil is small enough to see guests over.

BLUSHING BROMELIAD

NEOREGELIA CAROLINAE F. TRICOLOR

This impressive rosette-shaped bromeliad has a bright red centre and a cluster of small, lilac flowers, when mature. The leaves are patterned with bold green and yellow stripes. If you have a bright room with moderate or high humidity, this plant will thrive and create a dazzling display.

HEIGHT 30cm (12in)
SPREAD 60cm (24in)
FLOWERS Small, lilac
FOLIAGE Yellow-striped
LIGHT Filtered sun
TEMPERATURE 18–27°C (64–81°F)
CARE Fairly easy
PLACE OF ORIGIN Brazil

Funky foliage is ideal for a modern home or a very colourful interior.

CARE

The blushing bromeliad thrives in high humidity and is perfect for a bathroom. If you can commit to misting the leaves daily, you could place it in an area with lower humidity, such as a sitting room.

This plant doesn't have a particularly large root system, so a small container is suitable. In its native rainforest habitat, the blushing bromeliad grows in trees in a similar way to orchids, which explains why growing it in an orchid compost produces the best results.

Water the blushing bromeliad into the crown of the plant with distilled water or rainwater. Every few weeks, tip out water sitting in the crown and replace. Keep the compost just moist.

Mist the foliage regularly and apply a half-strength balanced fertilizer during the spring and summer.

After flowering for a couple of months, the plant will very slowly die. The good news is that it will then produce offshoots that will form the next generation.

PROBLEM SOLVING The most common problem faced when growing blushing bromeliads is stem rot, which is caused by overwatering.

New indoor gardeners may be unaware of the fact that bromeliads die after flowering – and it can come as a shock. For this reason, try to buy plants that are just about to flower and avoid any that are in full bloom.

DISPLAY

Display the blushing bromeliad so that you can see it from above, otherwise you'll miss out on the colourful drama that occurs during flowering: the centre of the plant turns red before the lilac flowers appear. A large, decorative bowl in the centre of a low coffee table would be ideal for this plant. Consider planting it with other bromeliads that look good when viewed from above, such as the urn plant (see p51) and scarlet star (see p.94). All thrive in orchid compost and benefit from regular misting.

The crown is where water collects in the blushing bromeliad's native setting.

ALSO TRY

Adventurous gardeners can create a bromeliad log. Remove bromeliads from their pots and place on a log that has hollows. Pack around the roots with damp orchid compost and tie in. Try these on your log:

- **Earth star** (*Cryptanthus bivittatus*), height 15cm (6in). This plant has wavy-edged leaves that are purple and pink in colour.
- **Zebra plant** (*Cryptanthus zonatus*), height 25cm (10in). This plant has leaves that are burgundy and cream in colour.

BOSTON FERN

NEPHROLEPIS EXALTATA 'BOSTONIENSIS'

One of the most popular of all houseplant ferns, the Boston fern is grown for its elegantly arching fronds and its classic look, which suits both traditional and modern interiors. If you're in search of a large but low-growing houseplant with a generous amount of foliage, this plant is ideal.

HEIGHT 90cm (36in)
SPREAD 90cm (36in)
FLOWERS None
FOLIAGE Arching
LIGHT Filtered sun/light shade
TEMPERATURE 12–24°C (54–75°F)
CARE Fairly easy
PLACE OF ORIGIN Central and South America and the Caribbean

CARE

The finely divided, arching fronds of this plant are at their best in a humid bathroom – frosted bathrooom windows filter the light to give the ideal growing enviroment for these elegant plants. There's a common misconception that all ferns like deep shade – in fact, filtered sunlight is preferred. Avoid a south-facing room with direct sunlight for the Boston fern as this will turn the foliage brown.

This shade-tolerant fern dislikes drying out, so keep an eye on watering.

Plant in a multipurpose compost and keep moist (but not wet); drying out can prove fatal to this plant. If it seems a little on the dry side, immerse the pot in water and then allow it to drain on the draining board. Apply a half-strength balanced liquid fertilizer in spring through to autumn and mist the plant regularly to ensure healthy foliage and good-quality fronds.

Repot every other year, but avoid planting too deeply as this can lead to rotting of the crown.

PROBLEM SOLVING Don't be alarmed if you find raised spots on the underside of the fronds. These spots are the spores, not pests, so no action is necessary. If fronds turn brown it's likely to be due to the compost drying out or the humidity being too low.

DISPLAY

This fern can be displayed in a number of ways: in hanging baskets, on plant plinths, as a single specimen, or as part of a mixed display. If you're feeling really adventurous, try *kokedama*, a Japanese technique that involves suspending the roots of the plant in a mud ball that's been coated in moss (see *p.19*). Other striking plants that would make effective partners for the hanging fern include maidenhair fern (see *p.50*) and the spider plant (see *p.65*).

Rabbit's foot fern is displayed here in *kokedama* style (roots coated in moss).

ALSO TRY

Turning your bathroom into a tropical paradise is easy when you grow plants that thrive in high humidity. Consider partnering the Boston fern with these fabulous, moisture-loving plants:

- **Rabbit's foot fern** (*Humata tyermanii*), height 30cm (12in). This is smaller than the Boston fern and is more suitable for a bathroom windowsill. It has lacy fronds and furry roots that cling to the sides of the pot.
- **Spider orchid** (*Brassia*), height 1m (3ft 3in). The spider-shaped flowers of this plant have thin petals of yellow and green that are marked with maroon spots.

OLEANDER *NERIUM OLEANDER*

This shrubby looking, easygoing plant is grown primarily for its stunning, scented pink flowers that appear between June and October. Pink is the most popular flower colour, but white and red forms are also avalable. Oleander has woody stems and dark green, narrow leaves.

HEIGHT 1.8m (6ft)
SPREAD 1.2m (4ft)
FLOWERS Pink, red, or white, scented
FOLIAGE Narrow, green
LIGHT Full sun
TEMPERATURE 13–29°C (55–84°F)
CARE Fairly easy
PLACE OF ORIGIN Mediterranean
WARNING! All parts are toxic; gloves required when handling

This popular, shrubby plant evokes the Mediterranean.

CARE

The oleander thrives in mild to warm climates, which has made it popular as a houseplant in areas where temperatures are too low for it to flourish outside. It's fairly easy to grow inside if it has sunshine and low humidity – a south-facing room is ideal as it needs at least four hours of bright sunlight a day to initiate flowering. It will also thrive in a sunny garden in the summer. Once the flowers are over, move the plant to an unheated room in the house to rest until spring.

Plant oleander in a loam-based compost with some horticultural sand and perlite mixed in to improve the drainage. Keep the plant well-watered throughout the summer and apply a high-potash feed to encourage flowers. Watering should be reduced dramatically during the winter resting period.

After flowering, prune the plant back hard. Cutting back each stem by one third will encourage more flowers the following year. It's important to wear gloves when handling oleander as all parts of the plant are toxic.

PROBLEM SOLVING Both aphids and scale insects (see pp.46–47) can be an issue for oleander. Aphids excrete a sticky honeydew which can lead to sooty mould (see p.47) if untreated. If you see a cluster of aphids, wipe them off with a damp cloth.

DISPLAY

You'll need plenty of space to display this impressive plant as it can become quite large, which is why it's frequently grown in conservatories. The oleander will also quickly swamp any plants that are growing with it in the same container. For this reason, always group it together with other sun-loving plants that will also thrive in a large room or in a conservatory, but make sure they have their own containers. Perfect partners for oleander include the paper flower (see p.6i) and jasmine (see p.103).

The delicate scent of pink oleander blooms is a great addition to a room.

ALSO TRY

Like the oleander, these plants can be moved outside onto the patio on warm days to make the most of the summer sun:

- **Blue African lily** (*Agapanthus praecox*), height 60cm (24in). Large heads of attractive blue flowers appear on this plant during the summer months. It makes a stunning conservatory plant at that time of year.
- **Yesterday-today-and-tomorrow** (*Brunfelsia pauciflora*), height 1.5m (5ft). This plant is similar in shape to the oleander but has flat, blue flowers.

BEAD PLANT *NERTERA GRANADENSIS*

This neat, eye-catching little creeping plant makes a wonderful windowsill specimen for a sunny, humid room: a kitchen is ideal but avoid a south-facing aspect as direct sun damages the foliage. The bead plant is grown for its plump, orange berries, which resemble pinheads in a pincushion.

HEIGHT 8cm (3in)
SPREAD 20cm (8in)
FLOWERS Insignificant
FOLIAGE Tiny, green
LIGHT Filtered sun
TEMPERATURE 13–20°C (55–68°F)
CARE Fairly easy
PLACE OF ORIGIN Chile
WARNING! Berries are mildly toxic

CARE

This is a fun plant that takes up very little space. It grows best when kept at 13–20°C (55–68°F). Although it will survive in higher temperatures, the chances of autumn berries forming after flowering are quite slim.

The bead plant is shallow-rooted, so it will grow in a shallow container filled with multipurpose compost. The kitchen is the perfect place to keep the plant as you're unlikely to forget to water it. In spring and summer, keep the compost moist at all times but during the winter months, allow it to dry out before rewatering.

Apply a half-strength balanced fertilizer by using a mister once a month during spring and summer. Mist every week with water, or more often in a room that has low humidity.

The bead plant can be quite tricky to keep year after year and encourage into berry. For this reason, it's often grown as a temporary plant, bought in berry, and then discarded once the berries are finished.

PROBLEM SOLVING If you keep your plant for more than one year and it fails to produce berries after flowering, this is likely to be due to it having been placed in a room that's too hot.

Another common issue is browning leaves, which is due to overheating and the plant receiving too much direct sun.

DISPLAY

Treat yourself to three or more bead plants and keep them on the kitchen windowsill; in the evening, move them to the centre of the dining table for decoration. Plant them in colourful pots and you'll have quite a glamorous display. Other low-growing, small plants that would look great alongside bead plants as a temporary centrepiece are the mosaic plant (see *p.91*) and African violet (see *p.126*).

Buy more than one bead plant and be adventurous when displaying.

ALSO TRY

Very few houseplants produce berries, but when they do, they certainly grab attention. If berries appeal and you can be sure young children won't be able to eat them when your back is turned, then consider these options:

- **Christmas pepper** (*Capsicum annuum*), height 45cm (18in). Cone-shaped miniature peppers appear on this plant just in time for Christmas.
- **Jerusalem cherry** (*Solanum pseudocapsicum*), height 45cm (18in). This plant bears large orange and red berries during the winter months.

BUNNY EARS CACTUS

OPUNTIA MICRODASYS

Offering the classic, much-loved cactus shape, this plant is a great choice. Its flat stems are covered in clusters of tiny spines and the summer flowers are bright yellow. If you want a low-maintenance plant with novelty appeal, the bunny ears cactus should be at the top of your list.

HEIGHT 30cm (12in)
SPREAD 45cm (18in)
FLOWERS Yellow
FOLIAGE Flattened, green stems
LIGHT Sun/filtered sun
TEMPERATURE 10–30°C (50–86°F)
CARE Easy
PLACE OF ORIGIN Mexico
WARNING! Plants are covered in fine spines; gloves required when handling

A plant for life: buy a small specimen and it will stay with you for years.

CARE

The bunny ears cactus will flourish on a sunny, south-facing windowsill and a room with low humidity – a bedroom would be perfect.

Plant in a pot of cactus compost and water no more than once a week in spring and summer. During the winter, you can get away with watering as little as once a month. To get the best out of your cactus and encourage flowers, feed with a specialist cactus feed every two months in spring and summer. Repotting is required every other year: this is a prickly and potentially painful task, so wear leather gloves; alternatively, roll up the plant in paper to protect yourself while handling.

If you want to propagate the cactus, do so in June: put on gloves, remove a fully grown segment, and leave for three days to dry out. After this time, fill a pot with cactus compost; then, wearing gloves, push the bottom of the segment 2cm (¾in) into the compost; lightly water to encourage rooting.

PROBLEM SOLVING As with all desert cacti, overwatering can be fatal. Too much water will cause rot and the plant will die. Underwatering is unlikely to be an issue, so if you're unsure how much to give your plant, keep to a stingy watering regime until you get to know it better. Mealybugs can also be a problem for this cactus (see p.46).

DISPLAY

Place this plant in a low-traffic area where its prickly spines won't be touched. It looks completely at home with other cacti or succulents. Plant a group in a large but shallow bowl to create a mini desert scene with a few rocks and some sand or grit. Zebra cactus (see p.96) and living stone (see p.106) make excellent partner plants.

Pincushion euphorbia has dramatic but very prickly spines.

ALSO TRY

If you're craving that desert look and an indoor display that requires very little in the way of care, then consider growing this plant, which has similar attributes:

- **Pincushion euphorbia** (*Euphorbia enopla*), height 30cm (12in). This euphorbia resembles a cactus, with green branches that are covered in stout, red spines.

SCENTED-LEAVED PELARGONIUM *PELARGONIUM*

HEIGHT 40cm (16in)
SPREAD 25cm (10in)
FLOWERS Pink, white, or red
FOLIAGE Scented
LIGHT Sun/filtered sun
TEMPERATURE 7–25°C (45–77°F)
CARE Easy
PLACE OF ORIGIN South Africa

There are many types of pelargoniums – all can be grown as houseplants. Scented-leaved forms are among the most popular, with fragrances such as ginger, rose, and lemon; these also have a number of reliable flowers.

The scented-leaved 'Little Gem' has rose-perfumed blooms all summer.

CARE

Scented-leaved pelargoniums are happy inside or out and will flower from spring to the end of summer, if given a well-ventilated spot. They thrive in high humidity and bright sunlight and are well-suited to a south-facing windowsill or a sunny conservatory.

Plant in a multipurpose compost and water well throughout spring and summer; allow the compost to dry out slightly between watering. Feed every other week with a balanced liquid fertilizer in spring to prompt healthy growth, then switch to a tomato feed in early summer to encourage plenty of blooms. Once the flowers have faded, deadhead them to promote more.

In winter, move to a cooler room to encourage the plant to rest. The following spring, cut back by a third, repot, step up the watering, and move the plant back to a sunny spot.

PROBLEM SOLVING If overwatered or grown in a room with high humidity, botrytis (a fungus) can be an issue. This appears as grey mould on the foliage. If caught early, it can be solved quite easily by removing affected leaves, reducing watering, and moving the plant to a more ventilated room.

DISPLAY

The scented-leaved pelargonium is often grown by those keen to create a cottage-style interior – a row of pelargoniums in terracotta pots, positioned on a sunny windowsill, is a great addition to a room with floral, country-style fabrics and rustic furniture. There aren't many flowering houseplants that tolerate a south-facing spot so, instead, couple pelargoniums with other sun-lovers, such as aloe vera (see p.53) and the money plant (see p.71), which are both praised for their succulent foliage.

ALSO TRY

Pelargoniums come in many forms. If you've fallen for their charms, expand your collection with other types that have the same needs and flower for months:

- **Ivy-leaved pelargonium** (*Pelargonium peltatum*), height 40cm (16in). This type has dark green glossy leaves and a more trailing habit than scented-leaved.
- **Regal pelargonium** (*Pelargonium grandiflorum*), height 45cm (18in). The large, blowsy flowers of this striking plant come in a wonderful array of colours.

Flowers of regal pelargoniums are flashier than scented-leaved types.

RADIATOR PLANT

PEPEROMIA CAPERATA

The radiator plant is a neat, mound-forming, tropical plant grown for its attractive, deeply corrugated foliage that can either appear green or maroon, depending on how the light catches it. During the summer months, it produces long, thin spikes that hold tiny, cream flowers.

HEIGHT 25cm (10in)
SPREAD 25cm (10in)
FLOWERS Spikes of slender, cream blooms
FOLIAGE Red or green, textured
LIGHT Light shade
TEMPERATURE 15–24°C (59–75°F)
CARE Fairly easy
PLACE OF ORIGIN Brazil

CARE

The radiator plant's native home is a tropical forest floor, so it enjoys shade and high humidity, which makes it an ideal candidate for a bottle garden (see p.18) or a north-facing room or bathroom. If you give these plants the right environment and keep the temperature consistent, they'll be easy to grow. However, they're unable to tolerate sudden changes in light levels and temperature.

Plant in multipurpose compost and, from spring through to autumn, water it as soon as the compost starts to dry out. Also during this time, feed your radiator plant with a half-strength balanced fertilizer once a month and reduce the watering in winter.

To keep up the humidity around the plant, place the container on a tray of damp, expanded clay granules: the radiator plant responds better to this than to misting the foliage.

To increase your collection of radiator plants, take cuttings in spring and simply pop them into a jar of water and watch the roots grow.

PROBLEM SOLVING If plants are kept too chilly and wet in the winter, they're likely to suddenly shed their leaves. They also react badly to cold draughts and will very soon start dropping their leaves. Rotting leaves are a sign of overwatering, while wilting leaves are often caused by underwatering.

DISPLAY

Bottle gardens are a wonderful way to enjoy plants that flourish in a humid environment – and they also add an interesting and appealing feature to a room. The radiator plant is an excellent candidate for this type of display, as are the maidenhair fern (see p.50) and the mosaic plant (see p.91).

The radiator plant is a great desktop companion, thanks to its neat shape.

A bottle garden can create a humid environment in a dry room.

ALSO TRY

There's a wide range of peperomias to choose from, including those with smooth foliage, trailing habits, and variegated leaves. Kickstart your collection with this plant:

• **Baby rubber plant** (*Peperomia obtusifolia*), height 30cm (12in). This attractive plant has smooth, dark green leaves that have a succulent-like appearance.

MOTH ORCHID *PHALAENOPSIS*

HEIGHT 90cm (36in)
SPREAD 60cm (24in)
FLOWERS Vast range of colours
FOLIAGE Dark green
LIGHT Filtered sun
TEMPERATURE 16–27°C (61–81°F)
CARE Easy
PLACE OF ORIGIN Southeast Asia

Moth orchids are the most widely available and easy to grow of all orchids. They're loved for their tropical-looking blooms, which come in a multitude of colours and with exquisite markings. If grown in filtered light and moderate humidity, the plant's arching stems will produce eight or more flowers.

CARE

Moth orchids will thrive on an east- or west-facing windowsill, where they can enjoy filtered sunlight. The ideal temperature to encourage flowers on the plants is 21°C (70°F). Avoid draughts and temperature changes.

These orchids should always be planted into a clear pot as their roots benefit from receiving light. A specialist orchid compost is perfect. This should be kept evenly moist with tepid rainwater as the plants react badly to hard tap water. The most effective way to water the plant is to soak the container and then allow the water to drain out onto the draining board.

Encourage flowers by feeding the plants in spring and summer with a specialist orchid feed and by keeping them in a fairly tight pot. Overpotting will result in lots of foliage and no flowers, so only repot them when the moth orchid is pushing itself right out of the compost. Repot into a slightly larger container.

Once flowers have faded, cut the stem back to just above the second node from the base of the plant.

PROBLEM SOLVING Too much direct sunlight will result in the yellowing of leaves, and too little light will lead to very dark green foliage. For a plant to produce flowers and be healthy it should have light green foliage – so aim for this colour.

If the arching flower stems become limp and floppy, support them with a small, green garden cane and plant ties.

DISPLAY

For real success, moth orchids need to be planted in fairly tight, individual containers – for this reason, they're not suitable candidates for a mixed planting scheme. However, they look wonderful when they're grouped with other flowering plants such as the cymbidium (see p.75) and the rose of China (see p.98) – both these plants enjoy similar growing conditions to the moth orchid.

Allow light to reach the roots by growing orchids in a clear pot.

Filtered sun and high humidity are care needs of noble dendrobium.

ALSO TRY

Orchids are among the most striking of all flowers. The most fanatical gardeners build orchid houses for their collections, but these plants happily grow in the home. Add to your collection with these beauties:

- **Noble dendrobium** (*Dendrobium nobile*), height 60cm (24in). This plant has stems of gorgeous orchid flowers, usually in pink and white.
- **Pansy orchid** (*Miltoniopsis*), height 60cm (24in). This orchid has fragrant, pansy-like flowers. Its blooms appear in the spring or autumn, depending on the hybrid.

HORSEHEAD PHILODENDRON

PHILODENDRON BIPINNATIFIDUM

The horsehead philodendron is the most dramatic of this group of plants. Its dark green, divided leaves are generous in size and make an instant impact in a room. If given the right conditions and environment, this non-climbing philodendron will soon create a truly magnificent houseplant.

HEIGHT 3m (10ft)
SPREAD 3m (10ft)
FLOWERS Insignificant
FOLIAGE Dark green, divided
LIGHT Filtered sun
TEMPERATURE 16–24°C (61–75°F)
CARE Easy
PLACE OF ORIGIN South America
WARNING! Sap is toxic; gloves required

CARE

This plant, also known as the fiddle-leaf philodendron, takes up a significant amount of space, and given a large room with filtered light, it will quickly become a showstopper in your home. In its native habitat, the plant grows as a self-supporting tree and sends down long aerial roots that act as props and help the plant to hold its own weight. As it can become top-heavy when mature, plant it in a heavy container filled with loam-based compost. Keep the compost just moist throughout spring and feed with a balanced liquid fertilizer every month. Reduce the watering during the winter months.

The horsehead philodendron enjoys a moderately humid atmosphere, so a generous misting of the foliage will be beneficial, as will wiping the leaves with a damp cloth to ensure the plant stays healthy and puts on its best display.

PROBLEM SOLVING If your mature plant becomes too big, prune it in spring by removing leaves from the top and sides, as needed. Make the cut right against the main stem. Wear gloves for this task as the sap is toxic.

As this plant is large, it's not unusual for the leaves to be damaged by people passing – remove any damaged foliage.

DISPLAY

This dramatic plant is as wide as it is tall, so it can be used as an effective room divider or a living screen to break up an open-plan room. To create an instant jungle scene, position your horsehead philodendron with other plants that enjoy filtered light and moderate humidity. The ideal partners are the Swiss cheese plant (see *p.109*) and climbing sweetheart plant (see *p.120*). For an added jungle look, consider putting the plant pots into decorative bamboo or rattan over-containers.

New leaves of horsehead philodendron are pale green until they fully mature.

ALSO TRY

Try these philodendrons to achieve a similar jungle drama, but on a far more manageable scale:
- **Blushing philodendron** (*Philodendron erubescens* 'Red Emerald'), height 1.2m (4ft). This plant has glossy, heart-shaped leaves with red stems.
- **Xanadu philodendron** (*Philodendron xanadu*), height 1m (3ft 3in). This is a mini version of the horsehead philodendron.

The Xanadu philodendron offers a modern, contemporary look.

SWEETHEART PLANT

PHILODENDRON SCANDENS

HEIGHT 1.5m (5ft)
SPREAD 1.5m (5ft)
FLOWERS Unlikely to bloom indoors
FOLIAGE Heart-shaped
LIGHT Filtered sun/light shade
TEMPERATURE 16–24°C (61–75°F)
CARE Easy
PLACE OF ORIGIN South America
WARNING! All parts are toxic; gloves required when handling

The sweetheart plant is an undemanding and vigorous climbing plant that's grown for its large, striking, heart-shaped leaves. The stems that carry this lush, jungly foliage look equally impressive tumbling gracefully from a hanging basket or climbing up a moss pole.

For a tall display, train the sweetheart plant up a moss pole.

CARE

There are few plants as accommodating and versatile as the sweetheart plant. Out of all the philodendrons, it's the one that copes the best in light shade and will flourish in almost any room in the house. It will also happily climb an indoor trellis and cover a wall.

Place the sweetheart plant in a container of multipurpose compost. During spring and summer, keep the compost moist. Feed with a balanced liquid fertilizer once a month and reduce the watering as winter approaches. The sweetheart plant will benefit from a regular misting of water to increase the humidity around this jungle plant.

Prune the plant in spring to keep it to the size you prefer. This is also the time for repotting. Increase your group of sweetheart plants by taking cuttings in late spring. To do this, remove the tip of a stem that has both a baby leaf and signs of small roots forming on it. Place the cutting in a jar of water to root.

PROBLEM SOLVING If the light levels in your room are too low, the sweetheart plant's growth will start to become leggy and the leaves will be small and turn yellow. If the tips of the leaves are brown, this is most likely to be due to the fact that the surrounding air is too dry. Remedy this by either increase your misting of the plant or move it to a more humid room, such as a bathroom.

DISPLAY

One of the most effective ways of displaying a sweetheart plant is on a moss pole (see *left*). Another good option is to invest in a large indoor hanging basket and display it with other attractive trailing plants that will also thrive in filtered sunlight and moderate humidity. You can create a spectacular display by combining the striking green foliage of devil's ivy (see *p.84*) and the maroon and silver foliage of the silver inch plant (see *p.139*) with the jungly sweetheart plant.

ALSO TRY

Plants with love-heart shaped foliage are popular as romantic gifts. If you'd like more options, try:

- **Heart-leaved fern** (*Hemionitis arifolia*), height 35cm (14in). This plant is ideal for a bottle garden.
- **Sweetheart hoya** (*Hoya kerrii*), height 25cm (10in). This succulent matures into a larger trailing plant.

Single leaves from the sweetheart hoya are often used as gifts.

MINIATURE DATE PALM

PHOENIX ROEBELENII

If you're in search of a really classic palm, the miniature date palm is an excellent choice. Its textured stem, elegant fronds, and architectural silhouette make it an almost unbeatable statement in a home. To show it off, choose a large room with filtered sunlight and moderate humidity.

HEIGHT 1.8m (6ft)
SPREAD 1.5m (5ft)
FLOWERS Insignificant
FOLIAGE Feathery fronds
LIGHT Filtered sun/light shade
TEMPERATURE 10–24°C (50–75°F)
CARE Fairly easy
PLACE OF ORIGIN China

CARE

Plant this spectacular palm in a large container filled with loam-based compost and water it moderately in spring and summer, ensuring that you allow the compost to dry out a little on top before rewetting. During this time, feed monthly with a balanced liquid fertilizer, but stop in autumn and reduce the water applied over the cooler months. These attractive plants thrive on moderate humidity, so mist your plant weekly.

This palm won't require pruning, apart from the removal of the bottom leaves if they start to turn yellow. There's also no need to repot the plant, unless it starts to become root-bound. Instead, simply add a fresh layer of compost to the top of the container in spring. Mature plants might produce insignificant yellow flowers, which are followed by small, black fruits.

PROBLEM SOLVING If growing your date palm in a conservatory or a room that has low humidity and high levels of sunlight, scale insects and red spider mites (see p.47) could be an issue. Take an occasional look at the underside of the leaves, where these pests tend to live, and act quickly to remove them.

DISPLAY

This spectacular palm deserves to be the focal point of a room and should never be squashed into a corner. For real impact, place a pair of palms either side of a doorway. If you've invested in mature specimens with trunks, give them an underplanting of the rattlesnake plant (see p.93) or the prayer plant (see p.107). Both these plants enjoy moderate humidity and filtered sunlight, just like the palm.

A mature specimen will instantly change the look and feel of a room.

ALSO TRY

Other palms that add immediate impact, height, and stature to a room include:

- **Areca palm** (*Dypsis lutescens*), height 2m (6ft 6in). This is similar in looks to the kentia palm (see p.100) but prefers filtered sunlight.
- **Fishtail palm** (*Caryota mitis*), height 2.5m (8ft). The triangular foliage of this plant looks like a fishtail. It's more upright than the miniature date palm.

The fishtail palm requires filtered sunlight and moderate humidity.

CHINESE MONEY PLANT

PILEA PEPEROMIOIDES

HEIGHT 30cm (12in)
SPREAD 30cm (12in)
FLOWERS Insignificant
FOLIAGE Round
LIGHT Filtered sun
TEMPERATURE 15–24°C (59–75°F)
CARE Easy
PLACE OF ORIGIN Southwest China

This plant, also known as the missionary plant, is grown for its round, coin-shaped foliage that resembles the leaves of a waterlily. It thrives in filtered sun and moderate humidity and is easy to place, but once you find the right spot, don't move the plant as it dislikes sudden changes in conditions.

Find the right spot and stick to it and the Chinese money plant will produce a crop of strong, healthy leaves.

DISPLAY

The succulent-like leaves with their waxy and shiny finish will add life and interest to any room in the house. The Chinese money plant's neat habit and polished appearance look particularly impressive in the centre of a coffee table or at a breakfast bar in a room that has filtered light. You can either choose to grow it as a neatly rounded plant by turning the container frequently, or allow it to become more one-sided by positioning it in a corner of a room. Display it with plants that have dramatically contrasting foliage shapes such as the velvet plant (see p.95) and the Boston fern (see p.112).

CARE

Some indoor gardeners recommend a shady spot for the Chinese money plant, but for the best foliage, sunlight – but not direct sunlight – is required. Turn your plant a couple of times a week to prevent it from growing towards the window and spoiling its shape.

Plant the Chinese money plant in a multipurpose compost in a fairly large container as it's a fast grower. Water when the compost starts to dry throughout spring and summer. Plants will quickly droop if they need water. In autumn and winter, reduce the water but continue to mist the foliage.

Apply a half-strength balanced fertilizer once a month in spring and summer to encourage a healthy plant.

This generous plant will produce a continuous supply of offsets that can be easily removed. Place your offset in a jar of water and watch the roots grow.

PROBLEM SOLVING The plant's leaves sometimes curl at the edges. This isn't overly worrying, but may be a sign it's getting too much shade. Move to a slightly brighter position; if this doesn't solve the issue, move again to an even brighter spot. The plant won't respond well to a vastly different environment, so a gradual move is sensible.

ALSO TRY

Houseplants with flat, circular leaves are extremely useful for introducing a change in pace, shape, and texture into a mixed display of foliage plants. If you like the idea of coin-shaped leaves, try the following option:
- **Button fern** (*Pellaea rotundifolia*), height 30cm (12in). This plant enjoys the same conditions as the Chinese money plant and has small, button-shaped leaves and elegant, arching fronds that look good in a hanging basket.

STAGHORN FERN

PLATYCERIUM BIFURCATUM

In their native habitat, staghorn ferns grow on other plants and get moisture and nutrients from the air. This is a tough environment to replicate, making them challenging plants to grow. Choose a room with filtered sun and high humidity: some indoor gardeners even grow them in the shower.

HEIGHT 90cm (36in)
SPREAD 90cm (36in)
FLOWERS None
FOLIAGE Antler-shaped fronds
LIGHT Filtered sun
TEMPERATURE 10–24°C (50–75°F)
CARE Challenging
PLACE OF ORIGIN Southeast Asia

CARE

This fern has two types of fronds: some are shaped like antlers, while others are smaller and have the important role of covering, or protecting, the root crown.

Plants are bought in pots, but for real success should be mounted onto a piece of wood or bark as they're happier growing on their side. To do this, remove from the pot, flatten the rootball and place onto a piece of flat wood or bark; pack around the edges with damp moss. Secure the moss and root to the board with clear fishing wire. Now you're ready to hang your mounted fern on a tiled wall.

Be careful not to overwater your fern if you choose to grow it in a pot.

DID YOU KNOW? THE STAGHORN FERN LIVES HIGH IN THE TREES IN ITS NATIVE HOME, CATCHING WATER THAT DRIPS FROM OVERHEAD LEAVES.

Water from above with rainwater and allow it to dry out slightly between watering; reduce watering further in the winter. Don't drown your plant — it needs humidity, not soaking wet roots. Feed monthly from spring to autumn with a balanced fertilizer. Never remove smaller fronds that cover the roots if they turn brown: the plant will shed them when it's ready.

PROBLEM SOLVING The staghorn fern is relatively disease-free. It's far more likely to suffer poor health if it's overwatered, or if it's placed in too much shade. A common mistake is to assume that all ferns are shade-loving.

If the fern appears to be lopsided, this may be due to its position: place it in line with a window, not to the side, so that it doesn't bend towards the light.

DISPLAY

Good air circulation and high humidity are perfect for the staghorn fern, so a large bathroom is ideal. Create a tropical-jungle look by growing it with other plants that enjoy the same damp environment, such as maidenhair fern (see p.50) and bird's nest fern (see p.57).

The butterfly orchid likes to be watered with rainwater.

ALSO TRY

If you like the idea of growing houseplants that in their native home grow in trees (these are known as epiphytes), then here's another good option to mount in your bathroom:

• **Butterfly orchid** (*Oncidium*), height 60cm (24in). This attractive, dainty orchid produces dozens of tiny flowers and is also happy mounted onto bark.

CAPE LEADWORT

PLUMBAGO AURICULATA

This popular greenhouse and conservatory plant flowers from June to autumn and needs little care. Cape leadwort is a reliable climber with stunning blooms that can be trained to cover walls on wires or a trellis. Plants can be kept smaller by pruning and used as freestanding specimens.

HEIGHT 4m (13ft)
SPREAD 3m (10ft)
FLOWERS Sky-blue
FOLIAGE Plain green
LIGHT Sun
TEMPERATURE 8–24°C (46–75°F)
CARE Easy
PLACE OF ORIGIN South Africa

CARE

The vigorous Cape leadwort requires a bright room with moderate to low humidity. You'll need to sweep up fallen leaves and petals occasionally, so a room with a hard floor is best.

Plant in a generous container of multipurpose compost and water well in summer, but reduce the watering as lower winter temperatures arrive. In very hot weather, mist the plant first

Restrain cape leadwort by pruning hard in autumn and training over a hoop.

Sky-blue flowers are produced in abundance and offer an uplifting display.

thing in the morning to avoid scorching the leaves, and feed every other week with a balanced liquid plant food (these are greedy plants).

The key to real success and plenty of unscented blooms is pruning. In autumn, cut stems back hard – cut just above a node at the base of the plant to prompt new growth. As flowers only form on new growth, if you don't prune, you won't get blooms.

PROBLEM SOLVING As with most plants that grow in full sun, there's a possibility they'll suffer from the sun-loving red spider mites (see p.47).

If your room temperatures drop below 8°C (46°F), the leaves will fall in autumn. To avoid this happening, prune the Cape leadwort back hard and keep it in a warmer spot during the winter months.

DISPLAY

Cape leadwort can be grown as a climber in a sunny entrance porch, so that its stunning blue flowers will bloom overhead. Alternatively, you can clothe the walls of your sunny room with these flowering, sun-loving plants. The Cape leadwort is often grown with paper flower (see p.61) or the highly scented jasmine (see p.103). At some point during the summer they'll all be in flower at the same time, which will create a truly breathtaking display.

ALSO TRY

If you're keen to grow flowering climbers up supports or walls there are many other options. To ensure you have a lengthy and colourful show, consider opting for long-flowering types, such as:

• **Passion flower** (*Passiflora caerulea*), height 4m (13ft). The flowers of this climber are wonderfully intricate and grace the plant from June all the way through to October.

• **White Cape leadwort** (*Plumbago auriculata* f. *alba*), height 4m (13ft). This plant has the same needs as blue Cape leadwort, but offers pure white flowers.

BAMBOO PALM _RHAPIS EXCELSA_

Of all the palms, this is the one to select if you have a shady room with moderate humidity. It has bamboo-like stems and fanned, ribbed leaves that have unusual, blunt ends. When mature, it will produce panicles of tiny yellow flowers, but it's primarily grown for its attractive foliage.

HEIGHT 2m (6ft 6in)
SPREAD 2m (6ft 6in)
FLOWERS Small, yellow
FOLIAGE Divided
LIGHT Shade/light shade
TEMPERATURE 10–25°C (50–77°F)
CARE Easy
PLACE OF ORIGIN South China

The **bamboo palm** has a good crop of foliage so is useful as a screen in a room.

CARE

The bamboo palm is very different in appearance from the more classic-looking houseplant palms. Its habit is dense and upright, so it won't need as much space as the arching types.

This slow-growing plant will enjoy a generous container. Plant it in multipurpose compost and allow it to slightly dry out between each watering from spring to autumn. Reduce the watering during winter months. Be vigilant about watering regimes – if the plant's in a shady room, the compost will take far longer to dry out, so overwatering can often become an issue. Apply a balanced liquid fertilizer once a month during the spring and summer.

Keep the plant looking tidy by removing any imperfect leaves or snipping off brown tips with scissors. The leaves have a blunt end, so no one will notice where you've made the cut. Wipe the foliage with a damp cloth if it gets dusty; this will also help ensure the plant's pores are clear.

PROBLEM SOLVING As the bamboo palm is an easy plant to grow, it rarely suffers from pests and diseases. It's more likely to suffer from being placed in the wrong environment. Bright sunlight and very dry air will cause the tips of the leaves to turn brown. To resolve this, move the plant to a shadier spot, mist the foliage, and snip off any damaged leaf tips.

DISPLAY

Finding a plant that will thrive in a shady room isn't easy, so this palm is a real asset if you have such a space, and it makes an attractive stand-alone specimen. When mature, its silhouette is really striking. In a modern setting, displaying a matching pair is a popular look. In a very shady room, grow bamboo palm alongside the equally easy-to-grow cast iron plant (see p.56) and the prayer plant (see p.107).

ALSO TRY

If the look of this striking bamboo palm appeals and you have plenty of indoor space to accommodate large plants, then you could also consider growing:

• **Mexican fan palm** (_Washingtonia robusta_), height 6m (20ft). This palm is likely to reach around just 2m (6ft 6in) indoors. However, the large, fan-shaped leaves make a real statement. Some gardeners grow the palm outside in summer and bring it indoors for winter. Larger specimens have one central trunk.

The Mexican fan palm's stout trunk is a large part of its appeal.

AFRICAN VIOLET *SAINTPAULIA*

African violets are grown for their gorgeous flowers, which can last for months and come in a range of colours, including purple, red, white, and pink. Their fleshy leaves are covered in a downy fur. These small plants, with their neat, compact habit, are ideal for east- or west-facing windowsills.

HEIGHT 13cm (5in)
SPREAD 20cm (8in)
FLOWERS Small, five-petalled blooms in a range of colours
FOLIAGE Fleshy, furry
LIGHT Filtered sun
TEMPERATURE 18–24°C (64–75°F)
CARE Easy
PLACE OF ORIGIN Tanzania

African violets make the perfect centrepiece, thanks to their neat shape.

CARE

Place African violets in a draught-free room with filtered light and moderate humidity and you'll hardly ever be without flowers, especially if you deadhead any faded blooms.

Plant in a small pot of multipurpose compost and water from below (*see image, above right*). The African violet enjoys humid conditions, but its leaves suffer if misted or splashed. Place on a tray or pot saucer of moist, expanded clay granules. When the compost starts to dry out, water from below, but don't let the plant stand in water for long. Reduce water in winter.

Water can mark the plant's foliage, so it's best to water it from below (*see p.35*).

Feed with a balanced liquid fertilizer once a month throughout the spring and summer months.

Move the plant to a south-facing windowsill when temperatures and light levels drop. Return it to an east- or west-facing spot in the spring. Repot African violets every two or three years, and increase your collection of these plants by taking leaf cuttings in late winter (*see pp.38–39*).

PROBLEM SOLVING If the leaf stalks become elongated, this is probably due to the plant having had too much shade. However, if it's placed in a very bright spot, brown marks will soon appear on the foliage. In both cases, pinch off severely affected leaves and move the plant to a different spot.

DISPLAY

The African violet is hugely popular: its neat shape, virtually year-long flowers, and soft, attractive leaves make it a striking plant to display in any interior. It's low and neat enough not to steal light in a shady room if placed in a small window, making it a great choice for a windowsill display. Place a collection of the plants on a long tray and, if space permits, add a few mosaic plants (*see p.91*) and Cape primroses (*see p.136*).

ALSO TRY

Other plants that are small enough to add interest to a windowsill but don't block the light include:
- **Earth star** (*Cryptanthus bivittatus*), height 15cm (6in). Ideal for a south-facing windowsill, this rosette-forming bromeliad has attractive and colourful foliage.
- **Peacock moss** (*Selaginella uncinata*), height 5cm (2in). This creeping moss will thrive on a north-facing windowsill, but it needs a humid atmosphere. As its name suggests, its iridescent, bluish-green leaves resemble the colours of a peacock's feathers.

HAHN'S SANSEVIERIA

SANSEVIERIA TRIFASCIATA 'HAHNII'

Hahn's sansevieria is a neat, rosette-forming succulent with attractive, grey-green leaves. It grows well in either filtered sun or light shade and requires very little in the way of care. Position it in almost any room in the house, apart from a bathroom as it dislikes high humidity.

HEIGHT 20cm (8in)
SPREAD 25cm (10in)
FLOWERS Insignificant
FOLIAGE Succulent, grey-green
LIGHT Filtered sun/light shade
TEMPERATURE 15–24°C (59–75°F)
CARE Easy
PLACE OF ORIGIN West Africa
WARNING! All parts are toxic; gloves required when handling

CARE

Plant Hahn's sansevieria in cactus compost to ensure good drainage. In spring and summer, water moderately, allowing the compost to dry between waterings. Apply a half-strength balanced fertilizer once a month. During autumn and winter, reduce watering, stop feeding, and protect the plant from cold draughts.

Hahn's sansevieria has an appealing and distinctive architectural silhouette.

If Hahn's sansevieria is under stress or needs repotting, it will produce insignificant white flowers. There's no immediate need to cut off these blooms, but make a note to repot the plant in spring. When repotting, remove any offsets and pot them up to create new plants.

PROBLEM SOLVING Most of this plant's problems are due to having been given too much attention. Overwatering often leads to rotting roots, after which, the plant's decline is rapid. Reduce the risk of rot by growing it in cactus compost, which is very well drained; thereafter, be careful not to overwater.

DISPLAY

Its rosette form and subtle colouring make Hahn's sansevieria a popular choice for displaying in small rooms that have a modern feel. Each leaf is patterned with grey markings; the tips of the leaves end in a sharp point, giving it clear definition and a distinctive, architectural shape. It's also often cited as being effective in purifying the air in a room (see pp.10–11).

Hahn's sansevieria can look stylish displayed in a simple grey container all on its own. However, consider grouping it with other shade-tolerant plants for even greater impact – it will grow happily alongside dumb cane (see p.77) and variegated snake plant (see p.128), a larger relative of Hahn's sansevieria. Both these plants enjoy the same watering regime and compost as Hahn's sansevieria.

ALSO TRY

If you like Hahn's sansevieria, you might also want to try its variegated form, which requires exactly the same conditions – together they make the perfect pair:

- **Golden bird's nest** (*Sansevieria trifasciata* 'Golden Hahnii'), height 20cm (8in). The thick leaves of this plant have a fine green marking around the edge, an outer golden band, and a green centre.

Golden bird's nest is an impressive plant when viewed from above.

VARIEGATED SNAKE PLANT

SANSEVIERIA TRIFASCIATA VAR. *LAURENTII*

Also known as mother-in-law's tongue, this popular plant is admired for its neat, upright shape and sword-like foliage. Its distinctive form offers a bold contrast to many other foliage plants. Yellow margins on the leaves heighten the drama.

HEIGHT 75cm (30in)
SPREAD 30cm (12in)
FLOWERS Insignificant
FOLIAGE Green with a yellow trim
LIGHT Filtered sun/light shade
TEMPERATURE 15–24°C (59–75°F)
CARE Easy
PLACE OF ORIGIN West Africa
WARNING! All parts are toxic; gloves required when handling

CARE

The variegated snake plant is easy to care for and thrives on neglect, so it's an excellent choice for offices and people who are often away from home. It copes well with low humidity and partial shade and is comfortable at the back of a room, where many other plants would suffer from insufficient light.

The variegated snake plant is an upright plant that's ideal for a tight corner.

Allow the top of the compost to dry out before watering.

Plant in cactus compost, allowing the surface to dry out between waterings in the spring and summer. During this active growing time, apply a half-strength, balanced fertilizer once a month. It's vital you protect the plant from cold draughts in winter and reduce watering at this time.

There's no need to repot the plant every year. If it needs repotting or is under stress, it will produce insignificant white flowers, which are held on an elegant stem.

PROBLEM SOLVING Water with caution as overwatering can cause rotting roots and the complete collapse of the plant. Growing it in a cactus compost encourages good drainage and helps to avoid death by overwatering.

During the winter months, be very careful not to give the variegated snake plant too much water.

DISPLAY

This plant is an excellent choice if space is limited in your home and you want to introduce height and impact. Each leaf can extend to as much as 75cm (30in). The eye-catching plant is more than capable of putting on an impressive display all by itself. However, if you plan to grow it as part of a group, pair it with plants that have contrasting leaf shapes such as the caster oil plant (see p.86) and the rubber plant (see p.88). All these plants enjoy filtered sunlight and moderate humidity.

ALSO TRY

The variegated snake plant is popular for its distinctive, upright shape. If you're keen to grow plants with similarly impressive foliage, consider this succulent:
- **Spear sansevieria** (*Sansevieria cylindrica*), height 75cm (30in). The smooth, bolt-upright, cylindrical leaves of this sansevieria have an even more dramatic architectural shape than those of the variegated snake plant.

CREEPING SAXIFRAGE

SAXIFRAGA STOLONIFERA

Creeping saxifrage is a mat-forming plant that has dense cushions of scalloped, velvety foliage and a dainty trailing habit, which makes it an ideal candidate for hanging baskets. It produces lots of plantlets that can easily be established as new plants for friends and family.

HEIGHT 15cm (6in), trailing to 60cm (24in)
SPREAD 40cm (16in)
FLOWERS Small, white
FOLIAGE Green, scalloped, purple-backed
LIGHT Filtered sun
TEMPERATURE 6–24°C (43–75°F)
CARE Easy
PLACE OF ORIGIN China, Japan

CARE

Creeping saxifrage offers a soft, dainty look but is nevertheless robust and easy to grow. It thrives in a room with moderate humidity and filtered sunlight. Some gardeners remove the spikes of flowers so the plant can put all its energy into growing foliage; others prefer to keep the small, white blooms.

The plantlets are held on long, wiry stems, adding to the drama of the display.

Plant in multipurpose compost in a hanging basket or place on a windowsill or shelf so that the runners that carry plantlets can freely trail. There's no need to cut off the plantlets.

Creeping saxifrage enjoys damp compost, so water well in spring and summer. Reduce the watering in the winter months. Feed every other week during the growing season with a balanced liquid fertilizer.

If you want to root the plantlets, it's easiest to grow creeping saxifrage on a windowsill. Propagation is very straightforward: simply push plantlets from the runners (still attached to the parent plant) into a nearby, smaller pot of multipurpose compost; hold them in place with paperclips. Once rooted (a few weeks), cut free from the parent.

PROBLEM SOLVING These fast-growing groundcover plants will soon suffer if positioned in a bright, south-facing room. They're far better suited to an east- or west-facing aspect. If the plants look unhappy and unhealthy, repot in spring – they dislike being pot-bound. They're susceptible to mealybugs (see p.46), but this is quite rare.

DISPLAY

Creeping saxifrage is best displayed in the home in a hanging basket or on a pedestal or a high shelf so that its elegant, purple-backed leaves are clearly visible. African violets (see p.126) and Cape primroses (see p.136) make great partners if you want to add flowering plants to a windowsill display.

(see p.46)

ALSO TRY

If you have a collection of hanging baskets or a shelf that has a little more room, then why not consider growing a variegated form of the creeping saxifrage, such as:

- **Magic carpet saxifrage** (*Saxifraga* 'Tricolor' (*stolonifera*)), height/trail 24in (60cm). This plant has green leaves with a white and pink trim and pure white flowers during the summer.

Magic carpet saxifrage is great in a room with pastel shades.

UMBRELLA TREE

SCHEFFLERA ARBORICOLA

Scheffleras produce individual leaflets that radiate out from a central stem, much like an umbrella, hence their common name. Indoor specimens are grown for their variegated foliage. They are ideal in a sitting room that has filtered light or light shade and moderate humidity.

HEIGHT 1.5m (5ft)
SPREAD 1m (3ft 3in)
FLOWERS Insignificant
FOLIAGE Variegated, divided leaflets
LIGHT Filtered sun/light shade
TEMPERATURE 15–24°C (59–75°F)
CARE Easy
PLACE OF ORIGIN Taiwan
WARNING! All parts are toxic; gloves required when handling

The umbrella tree is often sold with its stems supported by a moss pole.

CARE

The umbrella tree has been a popular houseplant for decades as it's so easy to care for. It can tolerate light shade but prefers filtered sunlight. When it is growing in a shadier room, you can expect a leggier plant.

Plant in multipurpose compost and apply a half-strength, balanced fertilizer once a month from spring to autumn. Water liberally during this time, but then reduce watering in winter. The umbrella tree is a tropical plant that thrives on a misting of water once a week, especially when the central heating is turned on as this will very quickly dry out the air in a room.

In its native habitat, it will grow into a vast tree; indoors, it's unlikely to exceed 1.5m (5ft). The plant responds well to pruning. An annual prune in early spring will not only keep it to your preferred size but will encourage more dense, bushy growth. If you'd like a tall specimen, give your plant the support of a moss pole.

PROBLEM SOLVING If growing in a warm room, the plant will be susceptible to red spider mites (see p.47). Some gardeners therefore choose to move their umbrella plants outside onto a sheltered patio during the summer. This often reduces the chances of an attack from these pests, which thrive in an indoor environment.

DISPLAY

This umbrella tree is often grown as a specimen on its own, but it also makes an attractive focal point in a mixed planting for a lightly shaded room. It combines well with dumb cane (see p.77) and peace lily (see p.133). The umbrella tree will grow much faster than these partners, so keep it well pruned if it takes over or you'll be left with an unbalanced display.

ALSO TRY

If you're keen to create a tropical, jungly scene in your home and you're new to indoor gardening, consider this easy-to-grow plant:

• **Silver sword** (*Philodendron hastatum*), height 3m (10ft). Silver sword's arrow-shaped leaves are a great contrast to the umbrella plant in shape and colour. In some lights, the foliage has a metallic look. Like the umbrella tree, it enjoys the support of a moss pole.

Silver sword is a low-maintenance plant with a climbing habit.

MIND-YOUR-OWN-BUSINESS

SOLEIROLIA SOLEIROLII

HEIGHT 5cm (2in)
SPREAD 60cm (24in)
FLOWERS Tiny, pinkish-white
FOLIAGE Very small, green
LIGHT Filtered sun/light shade
TEMPERATURE 7–24°C (45–75°F)
CARE Fairly easy
PLACE OF ORIGIN The Mediterranean

This much-loved plant has a creeping habit. Plant it in a shallow dish and it will cover the compost and make a stylish centrepiece. This is not a plant that can be ignored, so it's the wrong choice if you often forget to water or are away a lot.

CARE

Mind-your-own-business, also known as baby's tears, is highly adaptable to different light levels and will thrive in a humid environment, so a bathroom or kitchen is ideal. Keep the compost evenly moist all year and mist the foliage regularly: placing it by the kitchen sink will mean you're less likely to forget it. This plant will soon die if left to dry out. Plant in a pot of multipurpose compost and feed once a month with a half-strength balanced fertilizer.

This plant's neat, mound-forming habit makes it perfect for the centre of a table.

You have to look closely to spot the plant's tiny pinkish-white flowers.

Keep this fast-growing plant neat by clipping off any overhanging foliage with scissors. To create a new plant, just peel off a small section, with some roots attached, and pot on.

PROBLEM SOLVING Although mind-your-own-business enjoys a humid atmosphere, it can't cope with being swamped by other, larger plants. The foliage will quickly turn yellow and die if the plant is overcrowded (a frequent problem when it's grown in a bottle garden). To avoid this, remove any larger, overhanging leaves or grow the plant in its own container.

DISPLAY

Mind-your-own-business has miniature leaves and pretty, pinkish-white flowers that are so small you'll barely notice them, but its spreading power is great. It's often grown under other plants, but you'll have far more success displaying it in an individual pot as the foliage is very soon damaged when it's covered by larger, neighbouring leaves. Display it in a separate pot in a humid bathroom or kitchen. Place other plants that enjoy high humidity alongside mind-your-own-business to create a striking show. Great partners for this plant include the maidenhair fern (see p.50) and the bird's nest fern (see p.57). All these plants will flourish in the filtered light that's often produced by a frosted bathroom window.

ALSO TRY

Plants with very tiny leaves offer a fantastic contrast to those with giant and more decorative foliage. Why not find room for a few more attractive houseplants with miniature leaves, such as:

- **Button fern** (*Pellaea rotundifolia*), height 30cm (12in). Tiny, button-shaped leaves are held on this plant's arching fronds. It enjoys light shade and moderate humidity.
- **String of pearls** (*Curio rowleyanus*), trail 90cm (36in). This plant has long stems that are decorated with round leaves. It's wonderful for a hanging basket in a sunny room.

COLEUS *SOLENOSTEMON SCUTELLARIOIDES*

Every gardener should have at least one coleus in their home. This group of plants offers immense variety and is outstanding for the spectacular markings and colours of the foliage. Inexpensive and easy to raise from seed, coleus are most often grown as annuals.

HEIGHT 60cm (24in)
SPREAD 30cm (12in)
FLOWERS Insignificant
FOLIAGE Varied, colourful
LIGHT Filtered sun
TEMPERATURE 15–24°C (59–75°F)
CARE Fairly easy
PLACE OF ORIGIN Southeast Asia

Coleus will thrive inside or out during the summer months.

CARE

Coleus are extremely useful plants that will flourish in a moderately humid room with filtered sunlight. They will also be quite comfortable in a bedding or container scheme outside during the summer months.

Grow the plant in a multipurpose compost and keep it moist in spring and summer. If grown from seed, mix some slow-release feed into the compost. When growing it in

With so many colours available, you'll find one to suit your interior.

a room that has low humidity, mist the plant once a week. To encourage a bushy plant, pinch out the growing tips (with fingers or scissors). Flowers will appear on coleus in summer, but they aren't showy. Plants are often discarded in the winter as they usually become quite leggy.

If you like coleus, consider growing the plant from seed or taking cuttings in summer (see pp.38–41).

PROBLEM SOLVING When grown as a houseplant, coleus is sometimes susceptible to whiteflies or mealybugs (see p.46). Very badly affected plants should be thrown away as soon as possible to prevent the pests from spreading to other houseplants.

In spring and summer, if plants look leggy and short on leaves, cut the stems back by one-third and move plants to a sunnier spot.

There is little that can be done for plants in this state as winter approaches, so throw them away.

DISPLAY

Coleus forms a neat, bushy plant with soft stems. Leaf colours include red, yellow, lime green, orange, and pink. As there's so much choice with coleus, indoor gardeners often buy more than one plant: a group of them in a sunny room creates an impressive, uplifting display. Add to the drama by placing some begonias (see p.59) and Cape primroses (see p.136) into the mix.

ALSO TRY

When a plant offers such colour and interest through its spring and summer foliage you hardly need the addition of flowers. Other plants that present equally excitingly patterned foliage include:

- **Angel's wings** (*Caladium*), height 63cm (25in). The large, paper-thin, arrow-shaped leaves of this plant are beautifully patterned in green, pink, red, or white. The foliage will die down during the winter.
- **False shamrock** (*Oxalis triangularis*), height 30cm (12in). This plant has two-tone, dark purple, triangular foliage in spring and summer. It's ideal for a lightly shaded, cool room.

PEACE LILY *SPATHIPHYLLUM WALLISII*

The peace lily is a much-loved houseplant, thanks to its spikes of tiny white blooms encased by a white spathe. This elegant plant will bring a touch of sophistication to a room with light shade and moderate humidity. Reputed to reduce air pollutants, it's a popular choice for offices.

HEIGHT 60cm (24in)
SPREAD 50cm (20in)
FLOWERS Spike of white blooms
FOLIAGE Glossy, dark green
LIGHT Filtered sun/light shade
TEMPERATURE 12–24°C (54–75°F)
CARE Easy
PLACE OF ORIGIN Southeast Asia
WARNING! All parts are toxic; gloves required when handling

CARE

When buying peace lilies, look for plants in bud so that you can enjoy the spectacular pure white blooms for as long as possible. Plant in a multipurpose compost and keep the compost moist in spring and summer. As autumn and winter approach, let the surface of the compost dry out before rewatering.

Place the peace lily on a tray of moist, expanded clay granules to increase humidity around the plant. Feed every other week in the growing season with a balanced liquid fertilizer. The arching leaves shoot from individual stems directly from the compost and are joined in spring and sometimes again in autumn by the flowers. Once the flowers have faded, cut off the flower stem right at the base of the plant. Keep the leaves healthy and looking glossy by wiping them with a damp cloth from time to time.

PROBLEM SOLVING The one thing that peace lilies will not tolerate is a cold draught. This is fatal for them. An early sign that a plant might be suffering is black or brown edges to the leaves. To remedy this, move the plant to a slightly brighter, less draughty spot and trim off the marked part of the leaf.

DISPLAY

When fresh, the flowers are whiter than white, but after a few days will turn slightly green. The contrast between the pure white flowers and the glossy, dark green foliage is what makes this plant so elegant. Peace lilies are a popular choice for a mixed planter of houseplants as the white flowers brighten up the darker foliage of other plants that are tolerant of light shade. Place in a container with a Chinese evergreen (see p.52) and a trailing prayer plant (see p.107).

Peace lily's elegant shape makes it the perfect plant for a bedside table.

The central spike holds the flowers; the spathe offers a pure white backdrop.

ALSO TRY

If the rooms in your house only have light shade, try this distinctive fern:
• Crocodile fern (*Microsorum musifolium* 'Crocodyllus'), height 60cm (24in). This plant has glossy fronds that are magnificently patterned like crocodile skin. It's ideal for a room with high humidity, such as a steamy kitchen or bathroom, but will need space for its fronds.

STEPHANOTIS

STEPHANOTIS FLORIBUNDA

In summer, you'll smell the fragrant, white blooms of stephanotis before you see the plant. With a support to clamber up, this spectacular climber will cover an entire wall. It can be kept small by pruning and by growing over a pot hoop, which will allow you to enjoy it as a tabletop plant.

HEIGHT 3m (10ft)
SPREAD 3m (10ft)
FLOWERS Fragrant, white
FOLIAGE Thick, green
LIGHT Filtered sun
TEMPERATURE 10–23°C (50–73°F)
CARE Fairly easy
PLACE OF ORIGIN Madagascar

CARE

This stunning plant will add a touch of luxury to a room that has filtered sunlight and moderate humidity. Its clusters of stunning tubular flowers are long-lasting and, if grown in the perfect spot, can adorn the plant from May until the beginning of October.

When space is limited, you can easily keep stephanotis neat and compact.

Don't place stephanotis in a very bright room as this will cause the flowers to fade prematurely.

Plant in a container of multipurpose compost, and keep it moist between spring and autumn. Reduce watering slightly in winter. Apply a high-potash feed every two weeks in spring and summer to encourage a steady flow of flowers. If your room is on the dry side, for best results, mist the leaves at least once a week.

Prune in early spring to keep the plant to a manageable size; this is also the perfect time to repot, if required.

PROBLEM SOLVING Stephanotis react badly to being moved around, especially when they're in flower. Sudden changes in temperature and light levels will give rise to an unhealthy, ailing plant.

Failure to flower is most likely to result from lack of humidity or from the plant being kept too warm in the winter. During winter, it benefits from being in a cooler room, of around 13–16°C (55–61°F).

DISPLAY

The combination of dark, thick, glossy leaves and pure white flowers forms an impressive display. This, together with the plant's magnificent perfume, makes a desirable plant. Stephanotis is equally striking if kept small and trained over a pot hoop. Display it with other trained climbers such as the paper flower (see p.61) or rose

grape (see p.108). Don't display it with other scented climbers such as jasmine (see p.103) as the combined fragrances will be overpowering.

ALSO TRY

Here are some other large climbers that respond well to being trained over a pot hoop and pruned to a more manageable size:

- **Red passionflower** (*Passiflora racemosa*), height 4m (13ft). This plant is happy in a south-, west-, or east-facing room. It has striking red blooms. Prune after flowering.
- **Variegated Natal ivy** (*Senecio macroglossus* 'Variegatus'), height 1.5m (5ft). Grown for its ivy-like leaves, this plant can be pruned at any time of year. It enjoys full sun.

The red passionflower offers showstopping summer blooms.

BIRD OF PARADISE

STRELITZIA REGINAE

As its name suggests, the striking orange and blue flowers of this plant really do resemble a bird of paradise. Its large, paddle-shaped leaves are presented on stems that emerge from soil level and neatly point upwards to the ceiling. This plant makes a stunning focal point in a south-facing room.

HEIGHT 90cm (36in)
SPREAD 60cm (24in)
FLOWERS Large, orange, and blue
FOLIAGE Green, paddle-shaped
LIGHT Sun/filtered sun
TEMPERATURE 12–24°C (54–75°F)
CARE Challenging
PLACE OF ORIGIN South Africa
WARNING! All parts are toxic; gloves required when handling

Keep bird of paradise in a fairly tight pot to encourage flowering.

CARE

Although tricky to grow, this plant is worth the extra effort, as the flowers are so luxurious. Choose a room with good summer ventilation, moderate humidity, and full sun. A south-facing sitting room with patio doors is perfect.

Plant in multipurpose compost with added grit to improve drainage. Water regularly from spring until the start of

It is the mature plants that produce flowers, so be patient.

autumn to keep the compost moist. Gradually reduce watering over winter. Feed every two weeks with a balanced liquid fertilizer and mist the leaves regularly throughout the spring and summer. Moderate humidity is essential for the bird of paradise, so sit your planter on a saucer of moist, expanded clay granules, otherwise the plant will suffer in centrally heated rooms.

Don't overpot as a tighter pot will encourage flowers. Young plants take a few years to flower, but mature plants will bloom any time between April and November. In spring, cut off any unsightly leaves at the base of the plant. To keep the bird of paradise tidy and encourage further flowers, remove faded blooms as they start to wither. Repot every other year in spring.

PROBLEM SOLVING If the leaves are looking poor, small, or have brown tips, this is probably caused by insufficient water or light, or by humidity in the room being too low. Adjust the plant's conditions to determine which issue is causing the problems.

DISPLAY

This plant is a showstopper and doesn't need the company of others to stand out in a display. It can look particularly striking in a bright orange pot. If you'd like to group it with other flowering houseplants, then the angel's trumpet (see p.62) and the rose of China (see p.98) are both happy in a south-facing room with moderate humidity.

ALSO TRY

If you love the idea of tropical flowers in your home but are short of space, consider buying a few flower stems from a florist shop. Other houseplant blooms that will last well in a vase in a cool room include:

- **Noble dendrobium** (*Dendrobium nobile*), height 60cm (24in). This plant's flowers grow all the way up the length of the stem. Many colours are available.
- **Siam tulip** (*Curcuma alismatifolia*), height 60cm (24in). The pink and violet flowers of this plant are carried on long stems.

CAPE PRIMROSE *STREPTOCARPUS*

The Cape primrose is a popular, inexpensive plant that's grown for its trumpet-shaped flowers that appear from May to the end of October in a wide assortment of colours. The rosette of wrinkled foliage is a perfect backdrop for the blooms. This plant is the ideal size for a windowsill.

HEIGHT 50cm (20in)
SPREAD 50cm (20in)
FLOWERS Wide range of colours
FOLIAGE Trumpet-shaped
LIGHT Filtered sun/light shade
TEMPERATURE 12–24°C (54–75°F)
CARE Fairly easy
PLACE OF ORIGIN Madagascar

Plant several Cape primroses in a bowl to create a fine display.

colder months of the year. Feed with a high-potash plant feed every fortnight between spring and autumn to encourage a steady flow of flowers.

After flowering, deadhead to keep plants looking tidy and encourage more blooms. In autumn, older leaves will start to die off, so remove these and a new crop of young, fresh foliage will replace them in spring. This is also the time to repot plants into a slightly larger container. You can take leaf cuttings to create more plants (see p.39).

PROBLEM SOLVING Mealybugs (see p.46) are attracted to the Cape primrose. Make a habit of checking your plants weekly for the white, fluffy bugs – pick them off as soon as you spot them. They usually lurk under the foliage or around the crown of the plant. Throw away badly infested plants.

As soon as flowers have faded, pinch them off and wait for more blooms.

CARE

An east- or west-facing windowsill is ideal for Cape primroses. They should be in a bright area but away from direct sunlight. Plant in a pot of multipurpose compost. The plants require moderate humidity, so sit the pot on a tray of moist clay granules to increase moisture content in the air. Water sparingly once a week between spring and autumn; allow the compost to dry out slightly between watering. Reduce watering during the winter – overwatering can lead to root rot. Move the plants to a south-facing windowsill for the

DISPLAY

Cape primroses are long-flowering and hugely rewarding houseplants that offer great value. They come in a wide range of sizes. Flower colours include blue, purple, pink, yellow, white, red, and bicoloured, so you're almost guaranteed to find a colour to suit your interior. Display individual containers of Cape primroses on a long windowsill tray filled with a layer of moist, expanded clay granules. Other plants that will combine well with them on a tray include begonias (see p.59) and African violets (see p.126).

ALSO TRY

If you're looking for other free-flowering houseplants that offer a wide choice of flower colour, you could try:
- **New Guinea impatiens** (*Impatiens* × *hawkeri*), height 50cm (20in). This is a large version of the popular bedding plant busy Lizzie. It flowers for months and enjoys a shady position in the home. Choose from white, pink, red, purple, and pink flowers.

STROMANTHE

STROMANTHE SANGUINEA 'TRIOSTAR'

The stromanthe's paddle-shaped foliage is exquisite. Each leaf is uniquely patterned with green and white and has a pink underside. These leaves are held on pink stems that grow up from the base of the plant. A bright, humid bathroom or kitchen is ideal for this stunner.

HEIGHT 45cm (18in)
SPREAD 60cm (24in)
FLOWERS Unlikely to bloom indoors
FOLIAGE Spear-shaped, variegated
LIGHT Filtered sun
TEMPERATURE 15–24°C (59–75°F)
CARE Fairly easy
PLACE OF ORIGIN Brazil

CARE

Plant in a pot of multipurpose compost and keep moist in spring and summer. Reduce the watering in winter. Make sure that the humidity levels around the plant remain high by misting all year and sitting the pot on a tray of moist, expanded clay granules. Feed with a half-strength balanced fertilizer every fortnight throughout the spring and summer months.

PROBLEM SOLVING Don't worry if the leaves roll up towards the central vein at night-time – this is quite normal. Stromanthe's foliage can scorch if it receives too much direct sunlight. Move to a shadier spot if leaves turn yellow.

DISPLAY

Some stromanthe leaves point directly upwards, while others grow to the side. With its haphazard growing habit and quirky leaf colour, this plant is both funky and stylish. Planted in an equally distinctive container, it will make a striking display in a humid room. The reddish flowers are unlikely to appear, but this is a plant that will hold attention with its foliage alone. Consider growing three stromanthe and planting each one in a different coloured pot (pink, green, and white) to reflect the foliage.

Stromanthe also looks distinctive grouped together with other plants. Accentuate the eye-catching pink in the leaves by growing it with other pink and maroon foliage plants such as the polka dot plant (see p.102) and radiator plant (see p.117).

ALSO TRY

Plants that have maroon, purple, or pink undersides to their leaves are very appealing – they look completely different according to the angle from which they're viewed. Another stunning plant with this distinctive feature is:
• **Boat lily** (*Tradescantia spathacea*), height 45cm (18in). This unusual, neat plant has attractive, deep maroon undersides to its thick, green, sword-shaped leaves. It thrives in a bright, humid spot – a bathroom or kitchen is ideal.

Mist stromanthe regularly, especially if growing in a centrally heated room.

AIR PLANT *TILLANDSIA*

Air plants are grown for their spiky or curled foliage and attractive, colourful blooms. These somewhat magical plants survive on fresh air alone – not a speck of compost is needed. They do, however, require a room with filtered sunlight and high humidity – a bathroom or steamy kitchen is ideal.

HEIGHT 10cm (4in)
SPREAD 45cm (18in)
FLOWERS Varied in colour and shape
FOLIAGE Silver-green, spiky
LIGHT Filtered sun
TEMPERATURE 15–24°C (59–75°F)
CARE Fairly easy
PLACE OF ORIGIN The Caribbean, Mexico, South America

Water air plants by placing them in a tray of tepid rainwater.

Display air plants by wiring them to an attractive piece of wood.

CARE

Air plants that are in a room with low humidity will need to be put in a tray of tepid rainwater for half an hour a week. Shake off excess water before returning to their original spot. If grown in a very humid setting, they won't need to be waterered so often. Mist plants regularly with rainwater and with *Tillandsia* fertilizer monthly. Misting or soaking the flowers can cause rot. Air plants only flower once in their lifetime and then die; blooms last about three weeks. Before dying, the plant produces offshoots that take its place.

PROBLEM SOLVING Overwatering can lead to rot, especially if plants are grown in enclosed glass containers.

This is easily spotted as the base of the plant will go soft and some of the lower leaves will drop off. If you catch the problem early enough, remedy by removing the lower leaves and making sure the air plants are left to completely dry out, before placing them back into their original position.

DISPLAY

Air plants are great fun to grow and display and make perfect gifts for novice gardeners. These small plants can be displayed in a number

> **DID YOU KNOW?** WATER AIR PLANTS WITH RAINWATER OR DISTILLED WATER: THE CHEMICALS IN TAP WATER HINDER THEIR ABSORPTION OF MOISTURE.

of interesting ways and are perfect plants for creative people. For example, experiment by growing them in large shells or clear, decorative bottles or other glass containers, or even attaching a collection of them to a piece of driftwood with wire (see pp.18–19). Have fun arranging air plants in different ways, but never use glue to attach them to an item.

The foliage is often more silver than green, and the flowers are surprisingly large for such a tiny plant and also very colourful, so choose plants with contrasting foliage shapes, sizes, and colours to combine with them.

Tillandsias are in the same family as bromeliads, so if you want to grow them alongside compatible plants, look to scarlet star (see p.94) and the blushing bromeliad (see p.111).

ALSO TRY

In their natural habitat, air plants live high in the trees and survive off the air and water that surrounds them. This type of plant is known as an epiphyte. Here's another striking plant from this category to try out:
- **Amazonian zebra plant** (*Aechmea chantinii*), height 60cm (24 in). The dark green leaves of this plant are decorated with a silver stripe. This bromeliad has red flowers surrounded by red, orange, and yellow bracts.

SILVER INCH PLANT

TRADESCANTIA ZEBRINA

HEIGHT 15cm (6in), trailing to 60cm (24in)
SPREAD 60cm (24in)
FLOWERS Small, purple
FOLIAGE Variegated
LIGHT Filtered sun
TEMPERATURE 12–24°C (54–75°F)
CARE Easy
PLACE OF ORIGIN Mexico

The silver inch plant is popular for its foliage, which has distinctive silver and green stripes and a purple underside – the silver shimmers in the light. This gently trailing, undemanding plant is simple to place in the home and makes a great addition to almost any interior.

CARE

Tradescantias enjoy moderate humidity and filtered sun. They will survive in a room with light shade, but the lack of light will cause the leaf colour to fade.

Plant in a pot or basket filled with multipurpose compost and water from spring to autumn so that the compost remains just moist. Reduce watering in winter. In a centrally heated room, mist weekly. Feed throughout spring and summer once a month with a balanced liquid fertilizer.

Pinch out the growing tips of plants in spring to promote new foliage and a fuller plant. The shoots you've removed as cuttings will root quickly in a jar of water (see *p.39*).

PROBLEM SOLVING Plants that are suffering because of too little light or water will produce leggy, leafless stems. This can be easily resolved by moving your plant to a brighter spot, watering more generously, and cutting back the spindly growth.

If a whole stem of leaves loses its variegation, simply cut this part of the plant away before the entire plant reverts to plain green.

DISPLAY

This plant looks at its best displayed in a hanging basket so that it can tumble elegantly over the side. It won't provide a long trail, but it will offer a neat, bushy display; prune at any time of year to keep the required shape. These aren't reliable flowerers, but it's a welcome treat when they do bloom. Either plant them in a hanging basket or place on a bookshelf or windowsill. If you're keen to fill an east- or west-facing windowsill with plants, display it alongside the African violet (see *p.126*) and Cape primrose (see *p.136*). All like moderate humidity, so place the pots together on a tray of moist, expanded clay granules.

The trailing habit of the silver inch plant makes it ideal for a window ledge.

The flowers of purple spiderwort appear sporadically through the year.

ALSO TRY

As they're so easy to care for, why not add a few more tradescantias to your collection? Here are two attractive, easy-to-grow relatives:

- **Purple spiderwort** (*Tradescantia pallida* 'Purpurea'), height 30cm (12in), trail 60cm (24in). This plant offers larger and longer leaves in dashing purple and small, pink blooms. It has a neat, trailing habit.
- **Small-leaf spiderwort** (*Tradescantia fluminensis*), height 20cm (8in), trail 60cm (24in). This is the most widely grown *Tradescantia*. It has smaller leaves than the silver inch plant.

SPINELESS YUCCA

YUCCA ELEPHANTIPES

Spineless yuccas range in size from small desktop plants to large, shoulder-height specimens. Bigger, more mature plants have an impressive palm-like trunk with a generous head of sword-shaped leaves at the top. This easygoing plant likes low humidity and is great for a sunny, south-facing room.

HEIGHT 1.5m (5ft)
SPREAD 75cm (30in)
FLOWERS Unlikely to bloom indoors
FOLIAGE Sword-like
LIGHT Sun/filtered sun
TEMPERATURE 10–27°C (50–81°F)
CARE Easy
PLACE OF ORIGIN Mexico

With its neat, upright habit, this yucca will offer height in a tight space.

CARE

Plant the spineless yucca in a large container in multipurpose compost. Large specimens can be top-heavy, so use your hands to make sure the compost around the plant is firmly in place. Let the top of the compost dry out between watering from spring to autumn. Reduce watering in winter. Apply a half-strength balanced fertilizer fortnightly in spring and summer.

In their native home, these plants grow into vast trees, but they're easy to keep to your preferred size – cut the stem to height with a handsaw in spring and it will resprout. Turn the plant regularly to prevent it from leaning towards the light. Repot every other year in spring.

PROBLEM SOLVING Don't worry if the lower leaves start to go brown. This is normal – simply peel or cut them off.

Aphids (see p.46) can be a problem with this yucca so watch out for them. If you spot them early, they can be easily wiped off with a damp cloth, avoiding the need to turn to a pesticide.

DID YOU KNOW? THE SPINELESS YUCCA WILL GROW HAPPILY OUTSIDE IN THE SUMMER. MOVE IT BACK INDOORS BEFORE THE COLD NIGHTS OF AUTUMN KICK IN.

DISPLAY

This is a plant that makes a statement. It's grown principally for its sword-like dark green foliage. If you display the spineless yucca against a white wall, its architectural shape and sharp, angular lines will create a dramatic shadow.

Larger specimens are most striking without partners as they offer a bold and flamboyant look all on their own. However, small specimens will combine happily in a mixed container with other undemanding sun-lovers such as the money plant (see p.71) and flaming Katy (see p.105). All three of these plants will flourish in low humidity and with minimal watering.

ALSO TRY

There are many different types of yucca and the majority are hardy and grown without protection in a garden setting. However, this is one other tender type that prefers being grown as a houseplant:

• **Spanish bayonet** (*Yucca aloifolia*), height 1.5m (5ft). This plant is almost identical in looks and care needs to the spineless yucca but it's not as neat in shape. It's often bought as a garden plant that needs protection over winter.

FERN ARUM
ZAMIOCULCAS ZAMIIFOLIA

This foliage plant has a unique appearance. The upright stems display pairs of very shiny dark green leaves along their entire length. If you're looking for a houseplant that's undemanding and will survive for long periods without water, the fern arum is ideal.

HEIGHT 90cm (36in)
SPREAD 60cm (24in)
FLOWERS Unlikely to bloom indoors
FOLIAGE Glossy
LIGHT Filtered sun/light shade
TEMPERATURE 15–24°C (59–75°F)
CARE Easy
PLACE OF ORIGIN East Africa
WARNING! All parts are toxic; gloves required when handling

CARE

The fern arum will thrive in a room that has filtered sun or light shade. It prefers low humidity, making it the perfect choice for a sitting room. It's a relative newcomer to the houseplant market and has shot to fame thanks to its impressive water-storing abilities. The fern arum stores water in its rhizomes (roots), which means that indoor gardeners can ignore the plant for weeks without it suffering any detrimental effects.

Thick, leathery leaves and firm stems make the fern arum a robust houseplant.

When repotting in spring, divide the fern arum to make more plants.

Plant in a 3:1 mix of multipurpose compost and horticultural sand. Water from spring to autumn only when the top of the compost has dried out. In winter, the plant may only need watering once a month. Feed with a balanced liquid fertilizer monthly during the spring and summer.

To increase your stock of plants, repot every other year and divide the parent plant (see p.41). It's at this point you can take a close look at its impressive, water-storing roots.

PROBLEM SOLVING Part of fern arum's popularity is that it seems to avoid any issues with pests and diseases – this is a bullet-proof plant that thrives on neglect. Overwatering is the most likely reason for it to ail. To avoid this, make sure the top 6cm (2½in) of compost is dry before rewatering.

DISPLAY

The fern arum is often described as having a primeval look as its leathery leaves and sturdy stems are so different from those of any other houseplant.

Finding a plant for a shady room with low humidity isn't easy. If you'd like to create a show with a group of plants in this setting, place the fern arum with other shade-lovers such as the equally easy-to-grow cast iron plant (see p.56) and the devil's ivy (see p.84).

ALSO TRY

If you're often away from home and would like some other undemanding houseplants that require little in the way of watering, consider growing:

- **Golden barrel cactus** (*Echinocactus grusonii*), height 50cm (20in). Often bought as a small specimen, this classic cactus is prickly, with a neat barrel shape. It's ideal for a sunny room.
- **Panda plant** (*Kalanchoe tomentosa*), height 60cm (24in). This succulent has velvety soft silver leaves and enjoys a sunny room with low humidity.

INDEX

Bold text indicates a main entry for the subject.

Author Tamsin Westhorpe

PUBLISHER ACKNOWLEDGMENTS

DK would like to thank Oreolu Grillo and Sophie State for early spread development for the series, Margaret McCormack for indexing, and Paul Reid, Marek Walisiewicz, and the Cobalt team for their hard work in putting this book together.

PICTURE CREDITS

The publisher would like to thank the following for their kind permission to reproduce their photographs:

Alamy Stock Photo: John Forsey 10tr; Kati Molin 13bc; Maritxu22 15tr; Sagar Simkhada 15bl; Eye Ubiquitous 16cl; Blue Jean Images 18tr; songyuth unkong 19cl; Prostock-studio 20cr; Cavan Images 22tr; Gulsina Shaina 24c; lermont51 26bc; Jonny Forsey 27tl; Nazarii Karkhut 27cr; Gina Kelly 27bc; BIOSPHOTO 30tr; Graham Turner 31cl; Maryna Mashkina 32tr; Dina Belenko 33tr; Piti Tantaweevongs 33c; Dmitry Marchenko 34tr; BIOSPHOTO 35bl; Liudmila Naumova 36bl; D. Hurst 37tc; BIOSPHOTO 37br; Gina Kelly 38tr; Arcaid Images 42tr; London Time 42bc; Pixel-shot 44br; Switlana Sonyashna 45tc; Nigel Cattlin 46tr; Art Directors & TRIP 46bc; Nigel Cattlin 47tc; WildPictures 47bl; Nigel Cattlin 47bc; Konstantin Malkov 50br; AY Images 57bc; hilmawan nurhatmadi 59br; Pixel-shot 64br; Deborah Vernon 65ca; R Ann Kautzky 66br; fotototo 68bl; Olga Miltsova 71bl; BIOSPHOTO 75c; A. Astes 78cl; Dorling Kindersley ltd 79bl; YAY Media AS 80c; ML Harris 81cr; WILDLIFE GmbH 87bl; Robyn Charnley 89cl; Universal Images Group North America LLC / DeAgostini 90cl; Glasshouse Images 91cr; MNS Photo 92cr; Dorling Kindersley ltd 95cl; blickwinkel 95bc; Panther Media GmbH 95br; EyeEm 96bl; Maritxu22 97br; Pixel-shot 100br; Maritxu22 102br; Yuval Helfman 103cr; Universal Images Group North America LLC / DeAgostini 104bl; Zoonar GmbH 105bl; Maritxu22 107cr; John Glover 110c; Oleksii Sergieiev 110bll; Elena Odareeva 110bl2; marek kasula 111cr; Fir Mamat 115cr; Chris Burrows 116cl; SCPRO 122cl; Avalon/Photoshot License 124c; Universal Images Group North America LLC / DeAgostini 125br; Mervi Veini 127bl; blickwinkel 129bl; Universal Images Group North America LLC / DeAgostini 130br; Olga Miltsova 131bl; Ian Redding 131tc; Jon Stokes 136cr; Stanislava Karagyozova 139cr; Fir Mamat 141c.

Dorling Kindersley: Mark Winwood / RHS Malvern Flower Show 59cl.

GAP Photos: Jacqui Dracup 43bl; Visions 58br; Visions 63bl; Visions 72bl; Friedrich Strauss 113cl; Nova Photo Graphik 119bl; Visions Premium 120cl; Visions 123bl.

Getty Images: Pekic 9bl; Westend61 10bl; KatarzynaBialasiewicz 11bl; mixetto 13tr; electravk 17bl; FollowTheFlow 20cl; staticnak1983 34bl; Helin Loik-Tomson 42br; magicflute002 68c; Photology1971 70c; Farhad Ibrahimzade 72cr; Westend61 82cr; serezniy 85cl; Tatyana Abramovich 98cl; FollowTheFlow 109br; emotionalsea 120br.

Illustrations by Cobalt id.

All other images © Dorling Kindersley

Produced for DK by
COBALT ID
www.cobaltid.co.uk

Managing Editor Marek Walisiewicz
Editor Diana Loxley
Managing Art Editor Paul Reid
Art Editor Darren Bland

DK LONDON

Project Editor Amy Slack
Managing Editor Ruth O'Rourke
Managing Art Editor Christine Keilty
Production Editor David Almond
Production Controller Stephanie McConnell
Senior Jacket Designer Nicola Powling
Jacket Co-ordinator Lucy Philpott
Consultant Gardening Publisher Chris Young
Art Director Maxine Pedliham
Publishing Directors Mary-Clare Jerram, Katie Cowan

First published in Great Britain in 2021 by
Dorling Kindersley Limited
DK, One Embassy Gardens, 8 Viaduct Gardens,
London, SW11 7BW

The authorised representative in the EEA is
Dorling Kindersley Verlag GmbH.
Arnulfstr. 124, 80636 Munich, Germany

A CIP catalogue record for this book
is available from the British Library.
ISBN: 978-0-2414-6020-7

Printed and bound in China

For the curious
www.dk.com